COMPUTER SCIENCE, TECHNOLOGY AND APPLICATIONS

VEHICULAR AD-HOC NETWORKS

APPLICATIONS AND TECHNOLOGY

COMPUTER SCIENCE, TECHNOLOGY AND APPLICATIONS

Additional books and e-books in this series can be found on Nova's website under the Series tab.

COMPUTER SCIENCE, TECHNOLOGY AND APPLICATIONS

VEHICULAR AD-HOC NETWORKS

APPLICATIONS AND TECHNOLOGY

G. G. MD. NAWAZ ALI
MD. NOOR-A-RAHIM
AND
PUSHPENDU KAR
EDITORS

nova
science publishers
New York

Library of Congress Cataloging-in-Publication Data

ISBN: 978-1-53618-038-1
Library of Congress Control Number: 2020938411

Published by Nova Science Publishers, Inc. † New York

CONTENTS

Preface		**vii**
Chapter 1	Role of Position-Based Routing Protocols for Different Applications of VANET *Biswa Ranjan Senapati and Pabitra Mohan Khilar*	**1**
Chapter 2	Elliptic Curve Cryptography in Intelligent Transport Systems *Muhammad Faisal Siddiqui, Muhammad Awais Javed and Rida Fatima*	**33**
Chapter 3	Exploitation of Network Coding in On-Demand Wireless Data Broadcast *G. G. Md. Nawaz Ali, Victor C. S. Lee, Yuxuan Meng, Peter H. J. Chong and Jun Chen*	**63**
Chapter 4	Fault Diagnosis Techniques for Mobile Wireless Networks *Hazim Jarrah, G. G. Md. Nawaz Ali and Peter H. J. Chong*	**113**

Chapter 5 Blockchain: Basics, Applications, Challenges and Opportunities **187**
Jaswant Arya, Arun Kumar, Akhilendra Pratap Singh, Tapas Kumar Mishra and Peter H J Chong

Chapter 6 Optimal Hole Coverage in the Presence of Dumb Node in Wireless Sensor Networks **239**
Pushpendu Kar

About the Editors **257**

Index **259**

PREFACE

From the past decade vehicular ad hoc networks got tremendous attention from the industry, academia and research community. According to US National Highway Traffic Safety Administration (NHTSA), there are more than 30 thousands fatalities caused by the vehicle accidents in the U.S. each year, which worth around $250 billion economic cost annually. Research shows that 82% of these accidents can be reduced by the successful deployment of vehicular networks, because nearly 75% percent of vehicular crashes are caused by inattentive drivers. Literally, vehicular ad hoc networks means a network forms by the vehicles. But it has been evolved to network with the infrastructure as well due to the inherent intermittent nature of vehicle-to-vehicle (V2V) connection. The high mobility of vehicles, wireless communication loss and range constraints are the main reason for this intermittent V2V connection. So, now vehicular networks means communication between vehicle-to-vehicle (V2V) and vehicle-to-infrastructure (V2I). There are billions of dollars invested to research, deployment and testing of vehicular networks. For the emerging connected and autonomous vehicle (CAV), a stable vehicular networks is the foremost requirement. It is now very much visible that CAV will be the future of Intelligent Transportation System (ITS). The book is dedicated to discuss for the techniques, applications and relevant technologies of vehicular ad hoc networks and its challenges. The first chapter discuss about the routing protocols of vehicular networks. It focuses on different position-based routing protocols and their mechanisms for the successful use of vehicular networks for different applications. The second chapter discusses on the security and privacy issues on vehicular networks. A well-known security technique called Elliptic Curve Cryptography (ECC) is discussed to secure vehicular data from various tampering attacks. The third chapter discusses on the on-demand wireless broadcasting mechanism for improving data dissemination performance in terms of data delivery ratio and response time.

A network-coding based approach has been investigated for improving the overall performance of existing classical data broadcast algorithms. The fourth chapter describes how to get a dependable system in the lossy communication medium. This chapter discusses on a number of fault diagnosis techniques, their strengths and weaknesses, and it reviews their implementations in mobile wireless networks. The fifth chapter discusses the basics of Blockchain technology, applications, research challenges and opportunities in the field. Finally, chapter six discuss about the identification and mitigation of the faulty nodes in the wireless network.

Chapter 1 - Vehicular Ad-hoc NETwork (VANET) is one of the subclasses of Mobile Ad-hoc Network (MANET). Two types of nodes are used in VANET. The vehicle acts as the mobile node and Road Side Unit (RSU) acts as the static node. On one side the number of vehicles is increasing day by day and on the other side availability of different sensors for various applications at affordable costs encourages the researchers to create a new ad-hoc network called VANET. The objective of VANET is to provide different Intelligent Transportation System (ITS) services to the end-users. The presence of a communication unit i.e., On Board Unit (OBU) in the vehicle and technological development of the electro-mechanical system changes the traditional carrier vehicles to smart vehicles. Due to the presence of the communication unit (OBU), the sensing unit, and the computational unit, VANET is used for different applications. Broadly, the applications of VANET are classified into four different spectra. These are safety-oriented applications, commercial applications, convenience applications, and productive applications. For the successful use of VANET, routing protocols play an important role. But designing a good routing protocol is a challenging task due to the characteristics of VANET. Unlike traditional MANET, the nodes in VANET move at greater speed. There is a frequent change in the topology due to the random motion of the vehicles at a different speed. In all regions, the density of vehicles is not uniform which may increase the vehicle data-carrying delay in the network. Also, performance in communication between the nodes is reduced due to the link instability between the nodes in the urban region because of the presence of high buildings and trees in the communication path. Also, the motion of the vehicles and the network capacity is constrained by the road size. Looking at the different applications and the constraints of VANET, routing protocols play a vital role. This chapter discussed different position-based routing protocols and their mechanism for the successful use of VANET for different applications.

Chapter 2 - The development of smart transportation systems all over the world is providing a lot of benefits and potential applications. Intelligent transport systems (ITS) aim to provide a safer, reliable and smarter transportation system with the help of wireless communications. Security and privacy are the major concerns when it comes to Intelligent Transport System, where the entire network is communicating wirelessly. Multiple cryptographic algorithms could be used to provide safe and reliable communication in ITS, however, the selected cryptographic algorithm should provide low latency. Elliptic Curve Cryptography (ECC) is a well-known technique to secure data from various tampering attacks. ECC uses Elliptic Curve Digital Signature Algorithm (ECDSA) for message signature and verification. The verification of messages containing ECDSA signatures is a lengthy and time-consuming process. ECDSA verification requires double scalar multiplication on the Elliptic curve points. For real time systems, the verification of such messages becomes a challenging task. To speed up the entire verification process, the authors focus on improving the double scalar multiplication time. In this chapter, endomorphism is exploited which divides the scalars into equal bit length parts. This reduces the range of data and parallel processing can be utilized efficiently. Reducing the range of data needs to be computed also reduces the memory utilization of the hardware; as the generated lookup tables are smaller in size. Furthermore, parallel processing provides faster computations. The proposed architecture is not fixed for a specific public key, thus making the system generic for any public key Q. The lookup tables are generated in real-time, thus providing better throughput. Results show that using multilevel endomorphism improves the total latency of the system by 2.3 times.

Chapter 3 - On-demand broadcast is a scalable approach to disseminating information to a large population of clients while satisfying dynamic needs of clients, for instance in vehicular networks. However, in conventional broadcast approaches, only one data item can be retrieved by clients in one broadcast tick. To further improve the efficiency of wireless bandwidth, in this work, the authors give a comprehensive study on incorporating network coding with representative on-demand scheduling algorithms while preserving their original scheduling criteria. In particular, a graph model is derived to maximize the coding benefit based on the clients' requested and cached data items. Further, the authors propose a heuristic coding based approach, which is applicable for all the on-demand scheduling algorithms with low computational complexity. The authors give a running example demonstrating how the proposed algorithm works for representative algorithms. In addition, based on various application requirements, the authors classify the existing on-demand scheduling algorithms into three groups, real-time, nonreal-time and stretch optimal. In view of different application-specific objectives, the authors implement the coding versions of representative algorithms in each group. Extensive simulation results conclusively demonstrate the superiority of coding versions of algorithms against their non-coding versions on achieving their respective scheduling objectives.

Chapter 4 - The development of dependable systems has always been of interest and challenge. Generally, the dependability of a system exhibits, to what extent; it can deliver correct services to the users, avoiding frequent and severe service failures. Mobile wireless networks such as MANETs (Mobile Ad Hoc Networks) and VANETS (Vehicular Ad Hoc

Networks) provide important services in critical situations. Hence, achieving the dependability in such systems is of most interest and challenging at the same time. This chapter, first, investigates the diagnosis requirements of mobile wireless networks. Also, it describes the main impairments of the dependability in such networks, including faults, errors, and failures. Second, it describes fault diagnosis as a powerful technique to cope with these impairments. Third, it discusses the most common fault diagnosis techniques that have been employed by fault-tolerant systems. Particularly, it describes these techniques, their strengths and weaknesses, and it reviews their implementations in mobile wireless networks.

Chapter 5 - Today, centralized systems are more common, and the reversibility of the process and transparency are kept hidden due to the ownership of a single authority. The consistency is difficult to maintain as the network is centralized, and the central server is responsible for managing overall traffic. Blockchain brings out the new features like decentralization, distribution, immutable, consistent and security by cryptographer hash algorithms. Blockchain technology is based on the consensus which hires distributed ledgers for enabling parties who do not fully trust each other to maintain a set of global states. The intermediaries work to maintain trust and the reduced number of intermediaries can make the process easier, faster, transparent and cheaper. The consensus mechanism is used to validate the transactions, and privacy is maintained by having multiple account addresses. This chapter presents the basics of Blockchain technology, applications, research challenges and opportunities in the field. As a recent phenomenon, Blockchain has the potential to bring a new perspective to security, resiliency and efficiency in the digital world. This chapter discusses each issue in details.

Chapter 6 - Wireless Sensor Networks (WSNs) are energy and resource constraint networks. So, activation of an optimum number of sensor nodes in a WSN is always desirable to increase the lifetime of the network. Sometimes, due to adverse environmental conditions, sensor nodes may exhibit dumb behavior. A dumb sensor node can continue its sensing operation but cannot communicate due to impairments from the adverse environmental conditions. The sensor nodes behave normally after the resumption of a favorable environmental condition. The dumb behavior of the sensor nodes results in a communication hole(s) in the network, which creates, increases, or decreases dynamically. The creation of hole(s) may lead to data loss and degrade network performance. In such a scenario, to resume network performance, the authors proposed *Dynamic Hole Coverage for Dumb Node (DHCDN)* scheme to mitigate the effects of dumb nodes from the WSN by activating the optimum number of sleeping sensor nodes in the network.

In: Vehicular Ad-Hoc Networks ISBN: 978-1-53618-038-1
Editors: G. G. Md. Nawaz Ali et al.

Chapter 1

ROLE OF POSITION-BASED ROUTING PROTOCOLS FOR DIFFERENT APPLICATIONS OF VANET

Biswa Ranjan Senapati **[*]** ***and Pabitra Mohan Khilar***
Department of Computer Science and Engineering,
National Institute of Technology, Rourkela, Odisha, India

ABSTRACT

Vehicular Ad-hoc NETwork (VANET) is one of the subclasses of Mobile Ad-hoc Network (MANET). Two types of nodes are used in VANET. The vehicle acts as the mobile node and Road Side Unit (RSU) acts as the static node. On one side the number of vehicles is increasing day by day and on the other side availability of different sensors for various applications at affordable costs encourages the researchers to create a new ad-hoc network called VANET. The objective of VANET is to provide different Intelligent Transportation System (ITS) services to the end-users. The presence of a communication unit i.e., On Board Unit (OBU) in the vehicle and technological development of the electro-mechanical system changes the traditional carrier vehicles to smart

* Corresponding Author's Email: biswa.rnjn@gmail.com.

vehicles. Due to the presence of the communication unit (OBU), the sensing unit, and the computational unit, VANET is used for different applications. Broadly, the applications of VANET are classified into four different spectra. These are safety-oriented applications, commercial applications, convenience applications, and productive applications.

For the successful use of VANET, routing protocols play an important role. But designing a good routing protocol is a challenging task due to the characteristics of VANET. Unlike traditional MANET, the nodes in VANET move at greater speed. There is a frequent change in the topology due to the random motion of the vehicles at a different speed. In all regions, the density of vehicles is not uniform which may increase the vehicle data-carrying delay in the network. Also, performance in communication between the nodes is reduced due to the link instability between the nodes in the urban region because of the presence of high buildings and trees in the communication path. Also, the motion of the vehicles and the network capacity is constrained by the road size.

Looking at the different applications and the constraints of VANET, routing protocols play a vital role. This chapter discussed different position-based routing protocols and their mechanism for the successful use of VANET for different applications.

Keywords: VANET, MANET, ITS, OBU

1. INTRODUCTION

Vehicular Ad-hoc NETwork (VANET), popularly known as the network on wheels, is one of the current growing research areas due to the technological advancement in sensing, computation, communication, wireless technologies, etc. [1-4]. For the luxurious life and to reduce transportation time, the demand for the vehicle is increasing day by day. Technological advancement tries to fulfill the demanding growth of people for vehicles. Figure 1 shows the number of vehicles manufactured in million units per year in India.

Also due to the advancement in the electro-mechanical-humanities department, various sensors are developed at affordable costs which are used for various wireless applications [5-7]. Thus, the availability of a large number of vehicles and sensors at reasonable prices creates a new network called VANET [8].

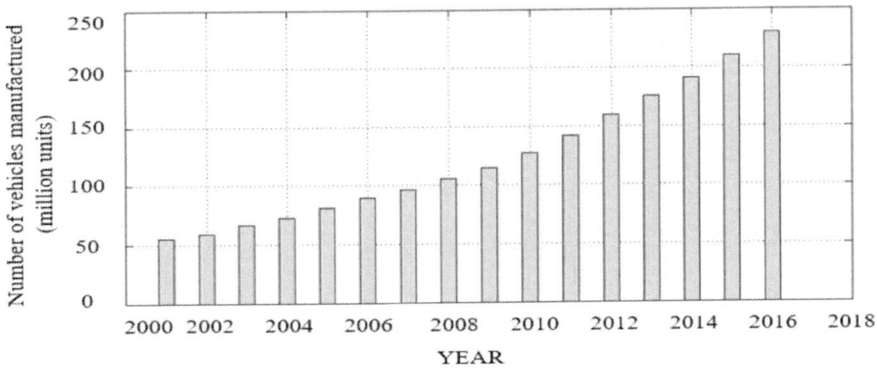

Figure 1. The number of vehicles manufactured per year.

Nowadays vehicles are not considered as a traditional carrier. Recent advancements in wireless technology, availability of Wi-Fi radio interfaces, GPS at low cost, and availability of various sensors at economic cost demands the development of smart and intelligent vehicles. Figure 2 shows the components of the smart and intelligent vehicles [9].

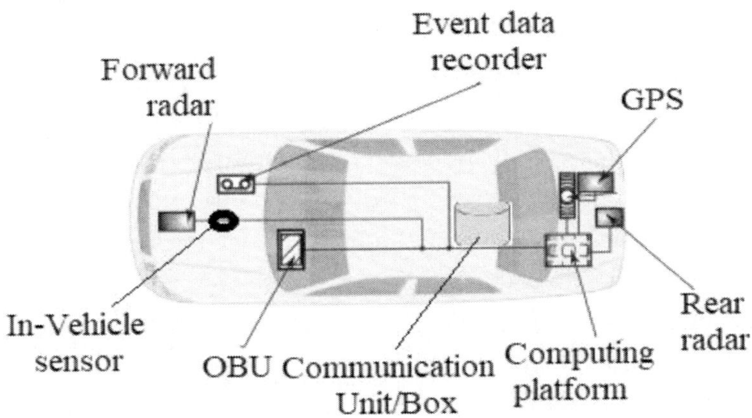

Figure 2. Components of a smart vehicle.

The smart and intelligent vehicles in these modern days are used for various applications. Figure 3 shows the different classification of applications of VANET.

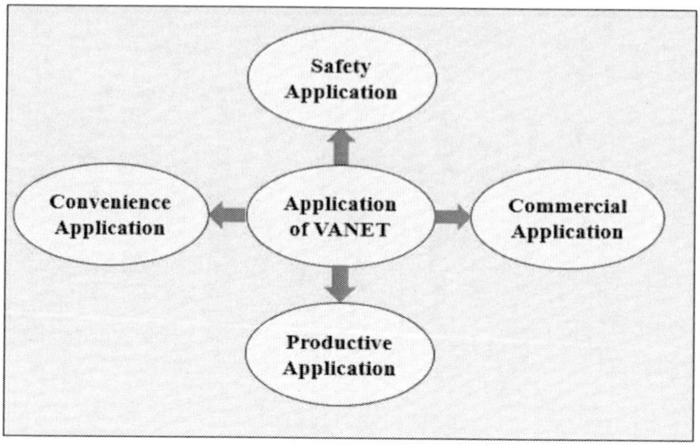

Figure 3. Classification of the application of VANET.

The primary objective of VANET is to provide safety applications. Safety application of VANET includes the provision of Intelligent Transportation Service (ITS) [10], reliable traffic data transmission for safety application [11], reduction in traffic jams by broadcasting the message to the neighboring vehicle [12], traffic congestion detection [13], etc. The convenience application of VANET includes automatic toll tax service through VANET [14], detection of emergency vehicles for faster movement in the heavy traffic [15], diversion of route in the heavy traffic [16], automatic parking slot detection [17], internet access through vehicles [18], etc. The commercial application of VANET includes buying and selling products through VANET [19], creation of virtual market through VANET [20], content sharing like advertisement through VANET [21], WiFi sharing through VANET [22], etc. The productive application of VANET includes environmental parameters monitoring through VANET [23], secure toll tax collection [24], etc.

VANET is considered as the subclass of Mobile Ad-hoc NETwork (MANET). VANET doesn't require any additional infrastructure for a new application is its main advantage. The presence of different sensors in the vehicle and the presence of computational and communication capability of the vehicle optimizes the cost of VANET for different applications. For the successful use of VANET for different applications, routing protocol plays a crucial role. The types of protocols frequently used in MANET is not directly used for VANET due to difference in characteristics between VANET and MANET. Table-1 shows the comparison of characteristics between VANET and MANET.

Table 1. Comparison of MANET and VANET

Characteristics	MANETs	VANETs
Network connectivity	High	Medium
Availability of energy	Low	High
Mobility models	Random	Restricted to road pattern
Speed of the node	Medium	High
Link Connectivity	Changes occasionally	Changes frequently
QoS	Low	High
Scalability	Medium	High

MANET generally prefers topology-based routing protocols due to the stable link connectivity between the nodes which changes occasionally. But due to frequent link disconnection and high speed of the node, topology-based routing protocols are not effective for the vehicular network. VANET generally prefers position-based routing protocols which help in the successful use of VANET for wide range applications. The major contributions of our survey chapter are as follows.

- Classification of different routing protocols used in VANET.
- Description of the working of the position based routing protocols with the corresponding advantages and disadvantages.
- Elaboration of the communication model used by different position-based routing protocols in VANET.

The remaining of this chapter is organized as follows. Section 2 mentions the Communication model for the vehicular network. Section 3 mentions the different categories of routing protocols used in VANET. Section 4 explains the techniques, benefits, and shortcomings of the position based routing protocols. Finally, the conclusion and future research scope are mentioned in Section 5. A list of abbreviations used for this survey chapter is mentioned in Table 2.

Table 2. List of notation and its meaning

Acronym	Meaning
VANETs	Vehicular Ad-hoc Networks
MANETs	Mobile Ad-hoc Networks
RSU	Road Side Unit
OBU	On Board Unit
V2V	Vehicle to Vehicle
V2I	Vehicle to Infrastructure
OSI	Open System Interconnection
WAVE	Wireless Access in Vehicular Environment
DSRC	Dedicated Short Range Communications

2. Communication Model for VANET

VANET consists of two important components.

- Vehicles – Act as the autonomous mobile nodes whose motion is restricted due to the structure of the road and can communicate with neighboring vehicles or RSUs present within its radio communication range.
- RSU – Road Side Unit which acts as the static nodes and the computational capacity, memory capacity, and the transmission range is higher than that of mobile node i.e., vehicles.

Based on these two components, communication in VANET is classified into two categories [25-27]. These are as follows.

- **Vehicle to Vehicle (V2V): -** Communication between the vehicles with another vehicle present within the communication range is called V2V.
- **Vehicle to Infrastructure (V2I): -** Communication between the vehicles and RSU or vice versa is called V2I or I2V communication.

Figure 4 represents the types of communication in VANET.

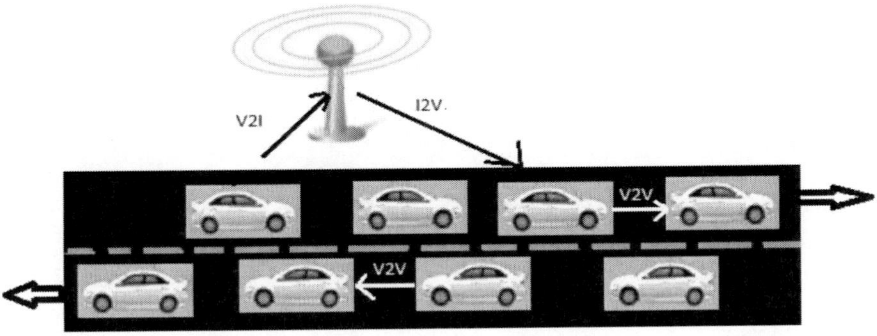

Figure 4. Types of communication in VANET.

Direct communication between the nodes (vehicles or RSUs) is possible if the nodes are present within the communication range. Otherwise, the communication between the nodes is multi-hop communication. To perform sensing, computation, and communication in a vehicle, an electronic part called On Board Unit (OBU) is attached to every vehicle. For the communication between the nodes, some standards are maintained in VANET. All the features of seven layers of Open System Interconnection (OSI) starting from the physical layer to the application layer are also maintained in the VANET standard to provide various features like medium access, security, etc. Two VANET standards are widely used for effective communication between the nodes. The first VANET standard is Wireless Access in Vehicular Environment (WAVE) [28].

Figure 5 shows the layered architecture of the WAVE model. WAVE standard defines the architecture, set of standardized protocols, services, and interfaces at all the WAVE stations to support the two types of communication in VANET i.e., V2V and V2I.

The second VANET standard is the Dedicated Short Range Communications (DSRC) [29]. DSRC is a standard developed by the USA which provides short to medium range communication between vehicles and RSUs [30].

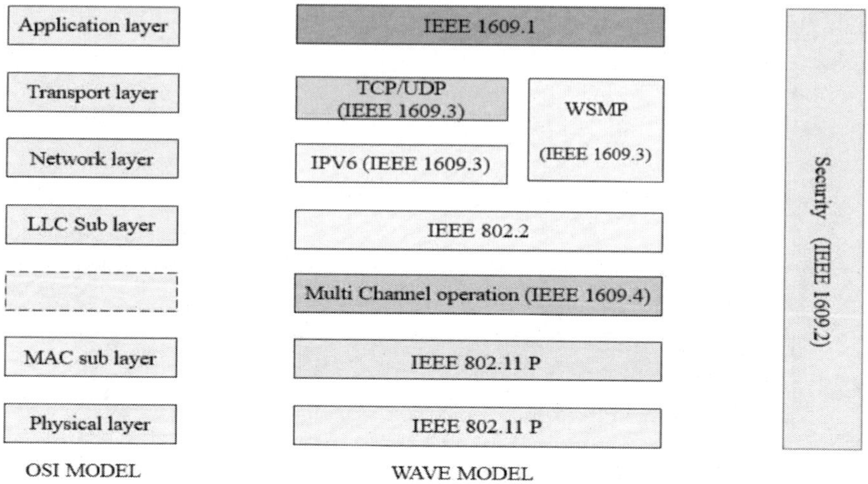

Figure 5. Comparison of the layered architecture of the OSI model with WAVE model.

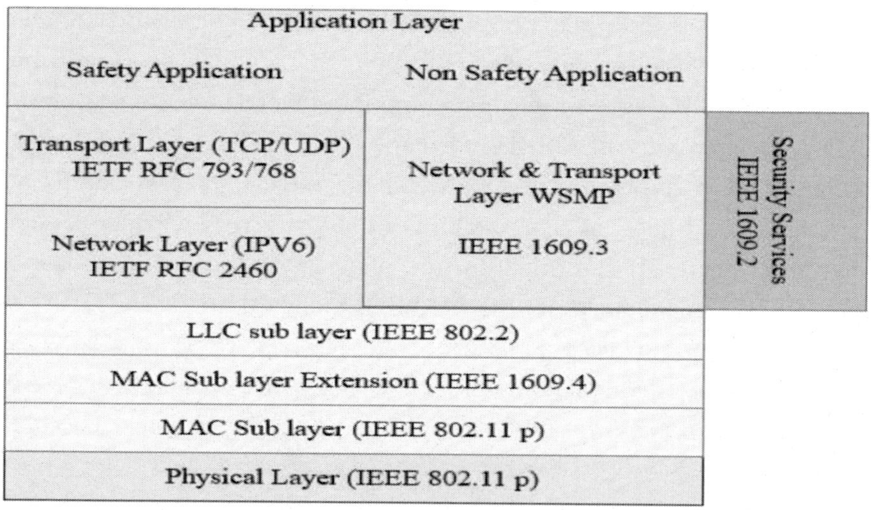

Figure 6. The layered architecture of DSRC model.

Both DSRC and WAVE standards are responsible for the effective communication between the nodes (Vehicles and RSUs) which is helpful to use VANET for the wide range of safety and non-safety applications [31]. For effective media access, the MAC protocol used by VANET is

IEEE 802.1p [32-33]. For the successful use of VANET, the role of routing protocol is crucial which is discussed in the next section.

3. ROUTING PROTOCOL FOR VANET

The objective of the routing protocol is to determine an efficient route from the source to destination to enhance the parameters of VANET like end-to-end delay, the number of hops, packet delivery ratio, etc. But the design of routing protocol is a challenging task for VANET. This is because of several characteristics of VANET like frequent link disconnection, random speed of the vehicle, the restricted motion of the vehicle due to the dependence on the road, the non-uniform density of vehicles in the network, restricted propagation model due to the presence of buildings and trees in the urban regions, etc. which are quite different from MANET. The routing protocols in VANET are classified into five categories which are shown in Figure 7 [34-35].

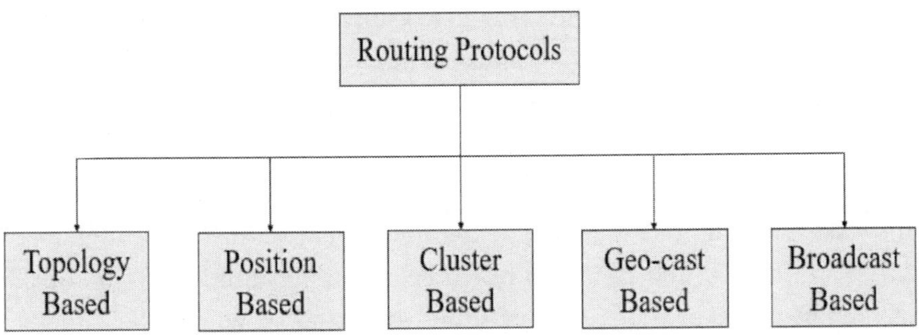

Figure 7. Types of routing protocol used in VANET.

3.1. Topology Based Routing Protocol

The link connectivity between the nodes is maintained for the forwarding of data packets between the nodes. Based on this link connectivity, this routing protocol is classified into three categories such as proactive routing protocols, reactive routing protocols, and hybrid routing protocols.

3.1.1. Proactive Routing Protocols

The proactive routing protocol is also called as the table-driven routing protocol. In this routing protocol, a link between all possible pairs of the node is maintained in the form of a table. The advantage of this protocol is that no route discovery process is required for the failure of a link as routes are already maintained. The disadvantage of this protocol is the wastage of memory for maintaining the table. This is because all the routes are not used during transmission from a source to destination. Unnecessary maintenance of routes that are not used in the routing led to the wastage of memory space. Examples of proactive routing protocols are Destination Sequenced Distance Vector (DSDV) routing [36], Optimized Link State Routing (OLSR) protocol [37], Fisheye State Routing (FSR) [38], etc.

3.1.2. Reactive Routing Protocols

The reactive routing protocol is also called as the On-Demand routing protocol. This is because it maintains the route information among the nodes which are currently in use. At the time of link failure or at the beginning of the communication, this protocol goes through a route discovery process to determine the link from the source to destination. Unnecessary wastage of memory is avoided because of the maintenance of required routes. At the link failure, routes are determined by flooding the route request packet that requires some time due to which the latency of the network increases. Examples of different reactive routing protocols are Ad-hoc On-demand Distance Vector (AODV) routing [39], Dynamic Source Routing (DSR) [40], Temporally Ordered Routing Algorithm (TORA) [41], etc.

3.1.3. Hybrid Routing Protocols

To overcome the limitation of proactive and reactive routing protocols, this protocol combines the features of the above routing protocols i.e., it can switch from proactive mode to reactive mode and vice versa. Different examples of hybrid routing protocols are Zone-based hierarchical link state Routing Protocol (ZRP) [42], Hybrid Ad-hoc Routing Protocol (HARP) [43], etc.

The overall category of topology-based routing protocol is shown in Figure 8.

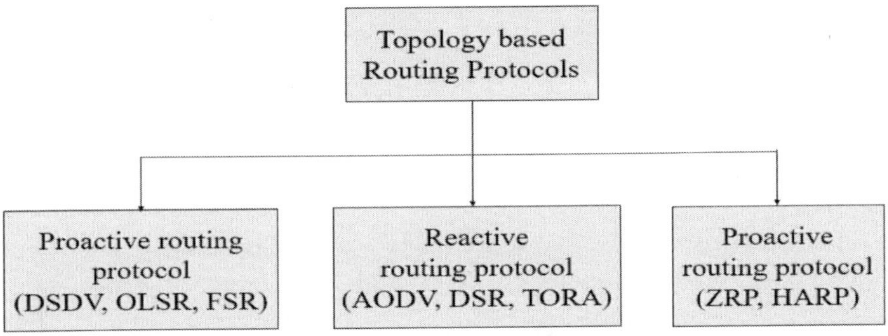

Figure 8. Different categories of the topology-based routing protocol.

3.2. Position-Based Routing Protocols

To overcome the limitation of topology-based routing protocols like frequent maintenance of routing tables, position-based routing protocols are used. The position-based routing protocol uses the geographic coordinates of a node to determine the route from the source node to the destination node. For this purpose, different location determination services are used like Geo Positioning System (GPS), DREAM Location Services (DLS), Reactive Location Services (RLS), Simple Location Services, etc. The advantage of this protocol is that it is suitable for VANET where there is fast mobility of the nodes. This type of routing protocol is more suitable for urban regions. The limitation of this protocol is the dependence on the location service to determine the position of a node for the establishment of the route. The overall structure of the position based routing protocol is shown in Figure 9.

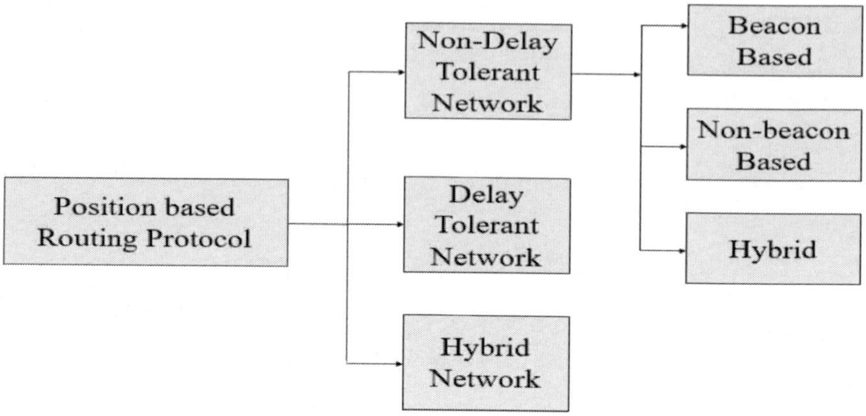

Figure 9. Different categories of the position-based routing protocol.

The brief description of each category of the position based routing protocol and corresponding advantage and disadvantage of the routing protocol is discussed in the next section.

3.3. Cluster-Based Routing Protocols

In the cluster-based routing protocol, the total network is divided into several small groups called clusters. In each cluster, a special node called cluster head is selected for the intracluster as well as for intercluster communication. This routing protocol is more scalable for a network having a high density of vehicles. At the same time, network overhead for this routing protocol increases due to the process of cluster formation and cluster head selection. Different examples of cluster-based routing protocols are Cluster-Based Routing (CBR) [44], Cluster-Based Location Routing (CBLR) [45], Hierarchical Cluster-Based Routing (HCBR) [46], etc.

3.4. Geo Cast Based Routing Protocols

Geo cast based routing protocol is the location-based multicast routing protocol. For this protocol, the entire network is divided into two regions. A Zone of Forwarding (ZOF) where the packets are forwarded without flooding and the Zone of Relevance (ZOR) where the packets are transmitted through flooding. Instead of using flooding for the entire network, flooding is used for a part of the network (ZOR) which reduces the network overhead. The division of the entire network into ZOR and ZOF is a challenging task. Different examples of Geo Cast based routing protocols are Inter-Vehicle Geocast (IVG) [47], Distributed Robust Geocast (DRG) [48], Dynamic Time Stable Geocast (DTSG) [49], etc.

3.5. Broadcast Based Routing Protocols

Broadcast based routing protocols are generally preferred for safety-oriented vehicles. This routing protocol mainly focuses on the communication between the vehicles (V2V) in a region having a high density of vehicles. This protocol is generally used to deliver advertisements in a particular location, broadcast road accident information, the occurrence of fire information, traffic information, road congestion information, etc. in the nearby region. The advantage of this protocol is the reliable transmission of data. But bandwidth consumption for this protocol will be more for a region having a high density of vehicles. Different examples of Broadcast based routing protocols are BROADCOMM [50], Urban Multihop Broadcast protocol (UMB) [51], Edge Aware Epidermic Protocol (EAEP) [52], etc.

A comparison of different routing protocols used in VANET is represented in Table 3.

Table 3. Comparison of different routing protocols for VANET

Protocols	Topology based protocol	Position based Protocol	Cluster-based Protocol	Geo Cast base protocol	Broadcast based protocol
Prior Forwarding method	Wireless multi-hop forwarding	Heuristic method	Wireless multi-hop forwarding	Wireless multi-hop forwarding	Wireless multi-hop forwarding
Digital map requirement	No	No	Yes	No	No
Virtual infrastructure requirement	No	No	Yes	No	No
Realistic traffic flow	Yes	Yes	No	Yes	Yes
Recovery Strategy	Multi-hop forwarding/ carry and forward	Carry and forward	Carry and forward	Flooding	Carry and forward
Scenario	Urban	Urban	Urban	Highway	Highway

4. Position-Based VANET Routing Protocol

Availability of different location-based service and different characteristics of VANET like frequent link disconnection, random speed of the vehicle, and restricted motion of vehicle support to use position-based routing protocol for VANET. Based on position-based routing, a network is classified into three categories like non-delay-tolerant network, delay-tolerant network, and hybrid network.

4.1. Non-Delay Tolerant Network

The non-delay-tolerant network is a computer network whose objective is to ensure the quick transmission of packets from source to destination without addressing any technical issues. Non-delay-tolerant networks classified the position-based routing protocol into three categories i.e., beacon-based, non-beacon-based, and hybrid.

4.1.1. Beacon Based Position-Based Routing Protocol

In this category of routing protocol, beacon signals are used to determine the distance between the vehicles. Beacon is a wireless device that is used to transmit radio waves continuously. Different beacon-based routing protocols are discussed as follows.

4.1.1.1. Greedy Perimeter Stateless Routing (GPSR)

The node which is far from the source node and nearer to the destination node and is present within the communication range is selected as the next intermediate hop for the further transmission in the GPSR protocol [53]. Two forwarding techniques – greedy forwarding and perimeter forwarding techniques are used to forward the packets to the destination. In the greedy forwarding techniques, the node which is at the optimum distance from the destination and presents far and within the communication range of the transmitting vehicle is selected as the next hop to forward the data. Due to the local maxima problem, the greedy forwarding technique fails. At the failure of greedy forwarding techniques, the perimeter forwarding technique is used which is based on the right-hand thumb rule. The main advantage of GPSR is the dynamic decision for the selection of the next forwarding node and the node remembers only one-hop neighbor information present within the communication range. The limitation of GPSR is that it is not suitable for the city environment due to the presence of a large number of obstacles like buildings, trees, etc. In the city region due to the frequent use of perimeter forwarding techniques as compared to the greedy forwarding techniques, the performance of GPSR in VANET is reduced. Figure 10 shows the two forwarding modes i.e., greedy forwarding mode and perimeter forwarding mode used in GPSR.

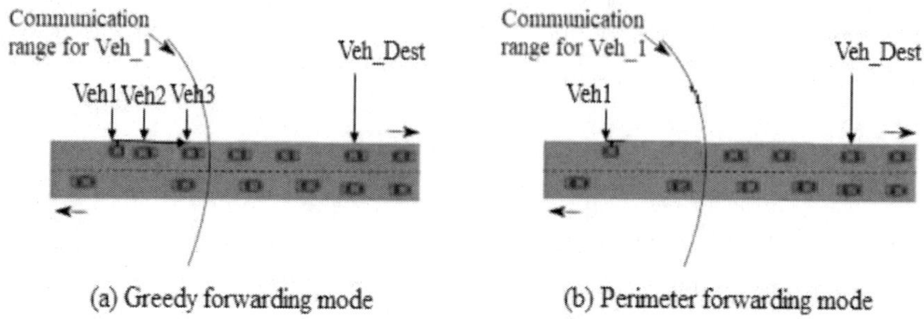

Figure 10. Greedy forwarding mode and perimeter forwarding mode of GPSR.

4.1.1.2. Geographic Source Routing (GSR)

A combination of position-based routing and the topological knowledge is used in GSR to avoid the limitation of GPSR [54]. The use of a static street map help to use GSR conveniently in the city environment. To determine the position of the destination node Reactive Location Service (RLS) is used which is based on Dijkstra's shortest path algorithm. Thus, the selection of a path from a junction towards the destination is dependent only on one parameter i.e., the shortest distance.

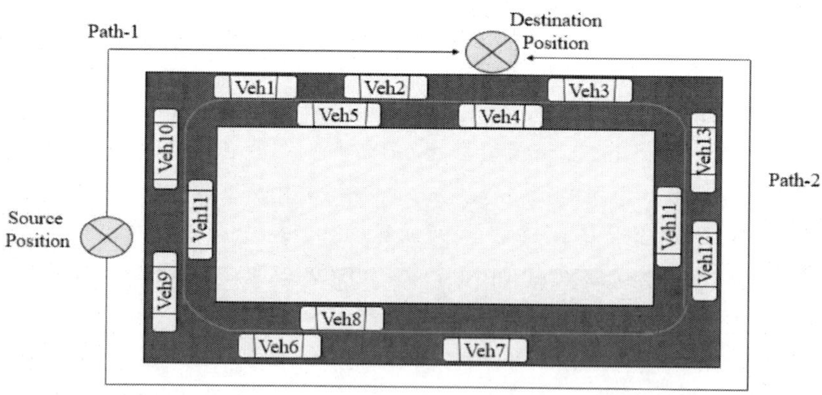

Figure 11. Selection of path using GSR.

Figure 11 shows a vehicular network. For the transmission of data from a source position to destination position using GSR, path-1 is selected based on the shortest distance as the parameter. GSR performs better in terms of total bandwidth consumptions and packet delivery ratio concerning some topology-based routing protocols like AODV and DSR. Considering the shortest path as the only parameter may reduce the performance of GSR if the network is sparse.

4.1.1.3. Anchor Based Street and Traffic-Aware Routing (A-STAR)

The pitfall of GSR is the selection of the next junction based on the shortest path without considering the connectivity between the nodes. This limitation is removed by a greedy location-based routing protocol called A-STAR which is suitable for the city environment [55]. For the transmission of a packet from source to destination, A-STAR uses vehicular traffic information based on regular periodic city bus routes. A-STAR computes the sequence of intermediate hops called anchors for the transmission of packets to the destination. The performance of A-STAR is better than GSR. Local recovery strategy using anchor node for A-STAR is more suitable for the city environment as compared to GSR and perimeter mode of GPSR. Due to the selection of the anchor nodes, most of the packets are transmitted through the fixed path based on connectivity. This increases the data congestion problem and also led to the lower packet delivery ratio. Figure 12 shows a vehicular network. From source position to destination position there are two paths (Path-1 and Path-2). To transmit the packet from source position to destination position, A-STAR selects Path-2 as the density of vehicles in Path-2 is more as compared to Path-1.

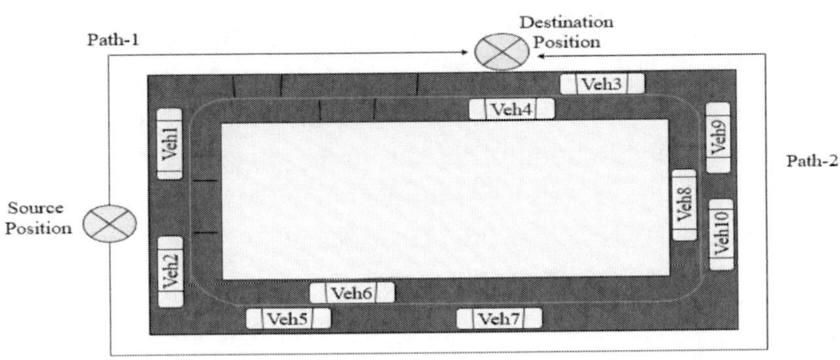

Figure 12. Selection of path using A-STAR.

4.1.1.4. Greedy Perimeter Coordinator Routing (GPCR)

The objective of GPCR is to remove the limitation of the perimeter mode of GPCR [56]. In the presence of radio obstacles like a tall building, a node present at the intersection is considered as the coordinator node even if the node is not closest to the destination. At the intersection, when the coordinator node receives the packet, then the further transmission of a packet from the coordinator node follows the greedy routing strategy to reach the destination. The selection of the coordinator node is important for GPCR and the selection is done by two approaches. The first approach is the neighbor table approach in which the position of the neighboring nodes is maintained and updated with the help of the beacon message. The second one is the correlation coefficient approach in which the coordinator node is selected based on the correlation coefficient which is calculated for every node with the help of the position of the node and the position of the immediate neighbor node. GPCR has the highest packet delivery rate than GPSR. Due to the use of the correlation coefficient approach, GPCR does not require any global or external information. The limitation of GPCR is the dependency on the coordinator node. But the coordinator selection fails at the intersection by the first approach for the curve road and by the second approach in a sparse network. Also, collision and packet loss occurs at the intersection by the coordinator nodes due to large packet transmission.

4.1.1.5. Connectivity Aware Routing protocol (CAR)

CAR protocol is suitable for the city environment and is based on the Preferred Group Broadcast (PGB) [57]. CAR uses AODV for route discovery and uses PGB for the broadcasting of data packet. After the determination of the location of the destination node and the route to the destination node, the packets are forwarded to the destination node with the greedy approach. Apart from this, CAR maintains the successful routes between different source and destination in cache memory. The information in the cache is updated periodically with the help of beacon messages. Even to reduce bandwidth consumption and network congestion, the beacon messages are piggybacked in the forwarded data packets. Two recovery strategies are used by CAR. The first one is Time Out with the active waiting cycle and the second one is the walk around error recovery. The benefit of CAR is no external information is required like a digital map. For the packet delivery ratio, the performance of CAR is better as compared to GPSR. This routing protocol faces difficulties when the traffic environment changes. The selection of anchor nodes also puts overhead on this protocol.

4.1.2. Non-Beacon Based Position-Based Routing Protocol

Non-beacon based position-based routing protocol does not use a beacon signal to determine the forwarder node. The determination of the forwarder node depends upon parameters like link connectivity, transmission and reception capability, etc. The explanation of a non-beacon-based routing protocol is discussed as follows.

4.1.2.1. Contention Based Forwarding (CBF)

Most of the beacon-based routing protocol uses the greedy forwarding mode approach. The forwarding mode may select a node with lesser connectivity which can reduce the end-to-end performance. The selection of the next road segment by considering the multi-hop connectivity rather than the vehicular density is the objective of TOPOCBF [58]. After the selection of the road segment, packets are forwarded using contention-based broadcast communication. This increases the packet forwarding reliability as compared to the unicast scheme. The performance of TOPOCBF is better in terms of packet delivery ratio and communication overhead. Elimination of a beacon message also reduces the bandwidth consumption. The performance of CBF is better in a network having high mobility of a node as it reduces the packet forwarding delay. The limitation of this protocol is that it is suitable only for the highway scenario rather than the city environment.

4.1.3. Hybrid Routing Protocol

As the name suggests, the hybrid routing protocol uses the approach of beacon-based mechanism and non-beacon-based mechanism to route a packet from source to destination. The explanation of a hybrid routing protocol is discussed in the next subsection.

4.1.3.1. Topology Assist Geo Opportunistic Routing (TO-GO)

The objective of TO-GO is to consider wireless channel error rates due to multipath fading and shadowing. TO-GO is a hybrid protocol as the selection of the next target forwarder is based on 2-hop beaconing and the quality of the wireless channel [59]. The design of the TO-GO protocol consists of various phases like next-hop prediction, forwarding set selection, priority scheduling, retransmission, and removal of a duplicate packet at the destination. It is based on two forwarding mode approach – greedy forwarding mode and junction forwarding mode.

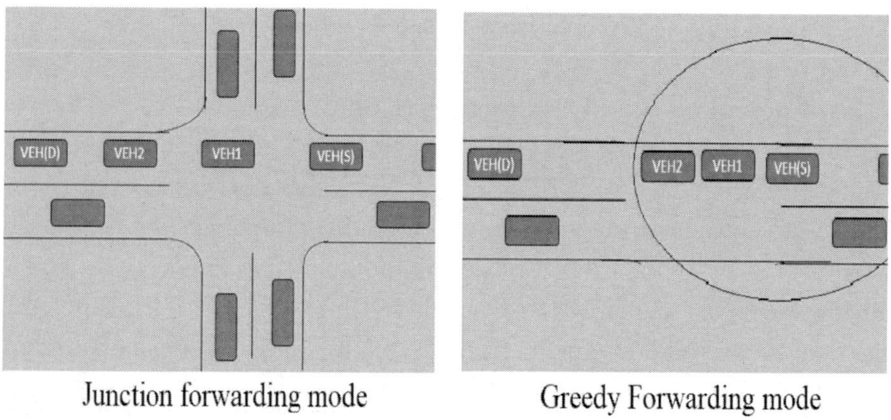

Figure 13. Two forwarding modes of TO-GO.

Figure 13 shows the two forwarding modes of TO-GO protocol. In Figure 13 VEH(S) is the source vehicle that wants to transmit the packet to the vehicle VEH (D). In the junction forwarding mode, instead of selecting VEH1, VEH2 is selected as the forwarder node based on two-hop information. In the greedy forwarding mode, VEH (S) selects VEH2 as the forwarding node. This is because VEH2 is far from VEH(S) and within its communication range. Also, VEH2 is nearer to the destination VEH (D). In TO-GO, the packet forwarding ratio is improved. This routing protocol avoids the hidden terminal problem. The limitation of TO-GO is the higher end-to-end delay as compared to GPSR and GPCR.

4.2. Delay Tolerant Network

Delay tolerant vehicular network is a network that addresses some technical issues like link disconnectivity by using store and carry forward method. The application of the delay-tolerant network involves asynchronous data traffic, loss of some data packets, etc. Explanation of some delay-tolerant vehicular network routing protocol is discussed in the next subsection.

4.2.1. Vehicle Assisted Data Delivery (VADD)

The objective of VADD is to minimize end-to-end delivery delays from a moving vehicle to a static destination in a sparse vehicular network [60]. The idea of carry and forward strategy is used by VADD. External information like digital maps and traffic statistics etc. are used by VADD. The working of VADD is based on three forwarding mode strategies. These are intersection mode, straightway mode, and destination mode. Figure 14 shows the three forwarding modes of VADD.

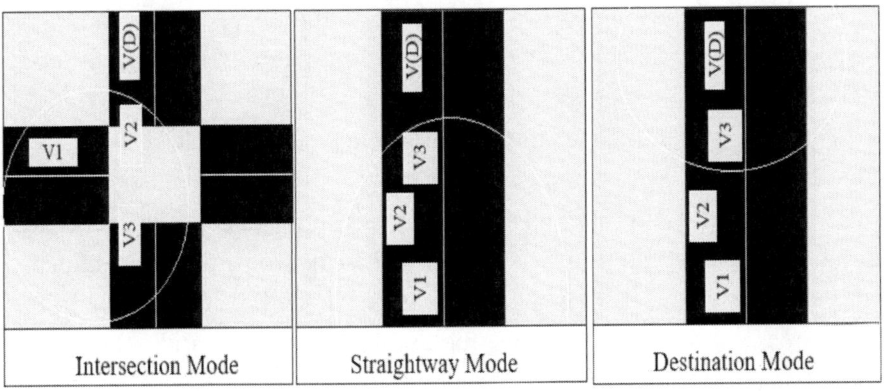

Figure 14. Forwarding modes of VADD.

In the intersection mode, the node determines the forwarder node moving in the direction of the destination node. In Figure 14, V2 and V3 are present at the intersection. Since V2 is moving towards the destination V (D) so V1 selects V2 as the forwarder node. The data forwarding in the straightway mode is simple as compared to the intersection mode due to bidirectional traffic. In this mode, the greedy approach is followed to select the forwarder node. From Figure 14, using the greedy approach V1 selects V3 as the forwarder node to transmit the data packet to destination V (D). In the destination mode, the location of the destination becomes the target location. When the packet reaches the target location, the GPSR approach is used. From Figure 14, the GPSR approach is used in the destination mode as V3 lies in the target location of V (D). The performance of VADD is better as compared to GPSR and GPCR in terms of packet delivery ratio, data packet delay, and traffic overhead. The performance of VADD is affected by the sudden change of traffic and topology.

4.2.2. Reactive Pseudo-Suboptimal-Path Selection Routing (RPS)

The objective of RPS is to select a path having maximum radio connectivity. Location management service and beacon message signals are generally used to determine the position and speed of a vehicle [61]. Vehicles are installed with a digital map to determine street information and traffic density. The working of RPS is based on three forwarding modes. These forwarding modes are intersection mode, segment mode, and RPS mode. Figure 15 shows the state transition diagram of different modes of RPS.

When the packet reaches the junction, intersection mode is used. In the intersection mode path having higher connectivity is selected. Segment mode is selected during the normal transmission of a packet in the straight path and the greedy approach is used for this method. In the straight path or at the intersection, if the packet forwarding node becomes isolated, then RPC mode is used. In the RPC mode, the forwarder node transmits a special packet to the previously traversed intersection to determine another path towards the destination. The performance of RPS is better in terms of packet delivery ratio, end-to-end delay as compared to GPSR. But due to

RPS mode, there is an increase in the overhead for the RPS routing protocol.

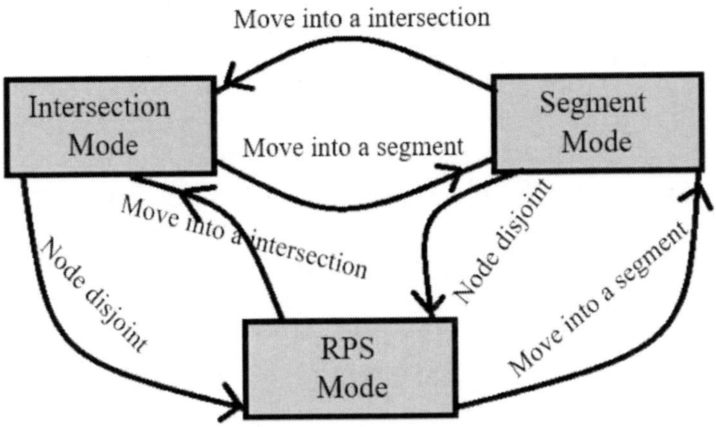

Figure 15. State diagram of different modes of RPS.

4.3. Hybrid Network

As the name suggests, more than one principle of position-based routing protocols is combined to form a hybrid network protocol. Different examples of hybrid network routing with its working mechanism are briefly discussed in the next subsection.

4.3.1. GeoDTN + NAV

It is a hybrid routing protocol as it is the combination of delay-tolerant and non-delay tolerant vehicular network routing protocol. Three modes are used for this routing protocol. These are the greedy mode, perimeter mode, and delay-tolerant vehicular network mode [62]. The ordering of the three modes used in this protocol is a greedy mode, perimeter mode, and at the end delay tolerant vehicular network mode. Figure 16 shows the state transition diagram of the three modes of GeoDTN + NAV. There is improved performance for GeoDTN + NAV in the sparse network scenario. The performance of GeoDTN + NAV is better as compared to GPCR and GPSR in terms of packet delivery ratio.

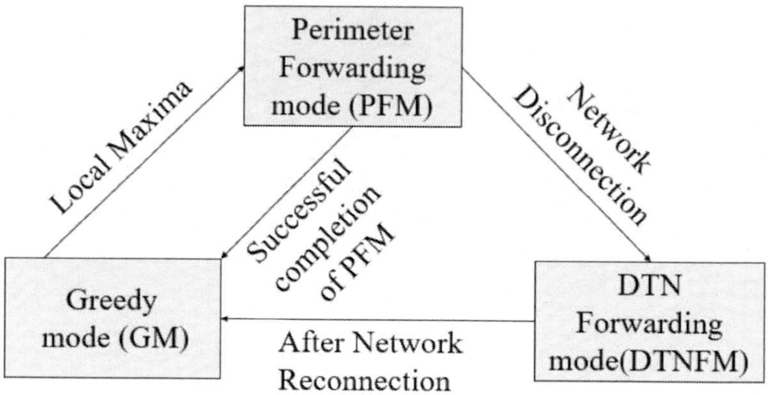

Figure 16. State transition diagram of different modes of GeoDTN + NAV.

4.3.2. Connectivity Aware Minimum Delay Geographic Routing Protocol (CMGR)

The objective of CMGR routing protocol is to improve the performance of the protocol in terms of packet delivery ratio in a sparse vehicular network. CMGR determines a path having maximum connectivity and low congestion. To determine the connectivity between the vehicles, beacon signals are used. To determine the congestion in a path, speed of the vehicle, etc. location service management GPS is used. The performance of CMGR is compared with A-STAR and VADD. For high vehicular density, the performance of CMGR is 25% better than VADD and A-STAR and for the low vehicular density, the performance is 900% better in terms of packet delivery ratio [63].

CONCLUSION AND FUTURE SCOPE

This chapter mentioned different position-based routing protocols for the effective communication between the nodes (Vehicles and RSUs) for the successful real-time application of VANET. Initially, VANET is compared with MANET. Different categories of application of VANET like safety-oriented applications, commercial-oriented applications, convenience-oriented applications, and productive applications are discussed. For the successful use of VANET for different applications, routing protocols in VANET are classified into five categories. Also, the comparison of the five types of routing protocols is done. This chapter finally reviewed different categories (Delay tolerant network, Non-delay tolerant network, and hybrid network) of position-based routing protocols. For the improved performance of the position based routing protocols, different external parameters are used like the vehicles are installed with digital map and location service management scheme to determine the road information, traffic condition, density of vehicles, etc. Analysis of different hybrid position-based routing protocols is the future research work. In the future, apart from position-based routing protocols, other categories of routing protocols like topology-based routing protocols, broadcast-based routing protocols, Geo-cast based routing protocols, and cluster-based routing protocols could be analyzed along with its merit and demerit.

REFERENCES

[1] Bhoi, Sourav Kumar, and Pabitra Mohan Khilar. "Vehicular communication: a survey." *IET networks* 3, no. 3 (2013): 204-217.

[2] Zeadally, Sherali, Ray Hunt, Yuh-Shyan Chen, Angela Irwin, and Aamir Hassan. "Vehicular ad hoc networks (VANETS): status, results, and challenges." *Telecommunication Systems* 50, no. 4 (2012): 217-241.

[3] Rasheed, Asim, Saira Gillani, Sana Ajmal, and Amir Qayyum. "Vehicular ad hoc network (VANET): A survey, challenges, and applications." In *Vehicular Ad-Hoc Networks for Smart Cities*, pp. 39-51. Springer, Singapore, 2017.

[4] Al-Sultan, Saif, Moath M. Al-Doori, Ali H. Al-Bayatti, and Hussien Zedan. "A comprehensive survey on vehicular ad hoc network." *Journal of network and computer applications* 37 (2014): 380-392.

[5] Mohapatra, Santoshinee, and Pabitra Mohan Khilar. "Artificial immune system based fault diagnosis in large wireless sensor network topology." In *TENCON 2017-2017 IEEE Region 10 Conference*, pp. 2687-2692. IEEE, 2017.

[6] Mohapatra, Santoshinee, and Pabitra Mohan Khilar. "Forest fire monitoring and detection of faulty nodes using wireless sensor network." In *2016 IEEE Region 10 Conference (TENCON)*, pp. 3232-3236. IEEE, 2016.

[7] Mohapatra, Santoshinee, Pabitra M. Khilar, and Rakesh R. Swain. "Fault diagnosis in wireless sensor network using clonal selection principle and probabilistic neural network approach." *International Journal of Communication Systems* (2019): e4138.

[8] Yousefi, Saleh, Mahmoud Siadat Mousavi, and Mahmood Fathy. "Vehicular ad hoc networks (VANETs): challenges and perspectives." In *2006 6th International Conference on ITS Telecommunications*, pp. 761-766. IEEE, 2006.

[9] Mejri, Mohamed Nidhal, Jalel Ben-Othman, and Mohamed Hamdi. "Survey on VANET security challenges and possible cryptographic solutions." *Vehicular Communications* 1, no. 2 (2014): 53-66.

[10] Khekare, Ganesh S., and Apeksha V. Sakhare. "A smart city framework for intelligent traffic system using VANET." In *2013 International Mutli-Conference on Automation, Computing, Communication, Control and Compressed Sensing (iMac4s)*, pp. 302-305. IEEE, 2013.

[11] Oliveira, Rene, Carlos Montez, Azzedine Boukerche, and Michelle S. Wangham. "Reliable data dissemination protocol for VANET traffic safety applications." *Ad Hoc Networks* 63 (2017): 30-44.

[12] Knorr, Florian, Daniel Baselt, Michael Schreckenberg, and Martin Mauve. "Reducing traffic jams via VANETs." *IEEE Transactions on Vehicular Technology* 61, no. 8 (2012): 3490-3498.

[13] Younes, Maram Bani, and Azzedine Boukerche. "Efficient traffic congestion detection protocol for next generation vanets." In *2013 IEEE international conference on communications (ICC)*, pp. 3764-3768. IEEE, 2013.

[14] Senapati, Biswa Ranjan, Pabitra Mohan Khilar, and Naba Krushna Sabat. "An automated toll gate system using vanet." In *2019 IEEE 1st International Conference on Energy, Systems and Information Processing (ICESIP)*, pp. 1-5. IEEE, 2019.

[15] Buchenscheit, Andreas, Florian Schaub, Frank Kargl, and Michael Weber. "A VANET-based emergency vehicle warning system." In *2009 IEEE Vehicular Networking Conference (VNC)*, pp. 1-8. IEEE, 2009.

[16] Jayapal, Cynthia, and S. Sujith Roy. "Road traffic congestion management using VANET." In *2016 International Conference on Advances in Human Machine Interaction (HMI)*, pp. 1-7. IEEE, 2016.

[17] Senapati, Biswa Ranjan, and Pabitra Mohan Khilar. "*Automatic parking service through VANET: A Convenience application*." (2019).

[18] Gerla, Mario, and Leonard Kleinrock. "Vehicular networks and the future of the mobile internet." *Computer Networks* 55, no. 2 (2011): 457-469.

[19] Bhoi, Sourav Kumar, Deepak Puthal, Pabitra Mohan Khilar, Joel JPC Rodrigues, Sanjaya Kumar Panda, and Laurence T. Yang. "Adaptive routing protocol for urban vehicular networks to support sellers and buyers on wheels." *Computer Networks* 142 (2018): 168-178.

[20] Lee, Uichin, Jiyeon Lee, Joon-Sang Park, and Mario Gerla. "FleaNet: A virtual market place on vehicular networks." *IEEE Transactions on Vehicular Technology* 59, no. 1 (2009): 344-355.

[21] Zhang, Yang, Jing Zhao, and Guohong Cao. "Roadcast: a popularity aware content sharing scheme in vanets." *ACM SIGMOBILE Mobile Computing and Communications Review* 13, no. 4 (2010): 1-14.

[22] Tufail, Ali, Mike Fraser, Ali Hammad, Kim Ki Hyung, and Seung-Wha Yoo. "An empirical study to analyze the feasibility of WIFI for

VANETs." In *2008 12th International Conference on Computer Supported Cooperative Work in Design*, pp. 553-558. IEEE, 2008.

[23] Senapati, Biswa Ranjan, Rakesh Ranjan Swain, and Pabitra Mohan Khilar. "Environmental monitoring under uncertainty using smart vehicular ad hoc network." In *Smart Intelligent Computing and Applications*, pp. 229-238. Springer, Singapore, 2020.

[24] Safi, Qamas Gul Khan, Senlin Luo, Limin Pan, Wangtong Liu, and Guangluo Yan. "Secure authentication framework for cloud-based toll payment message dissemination over ubiquitous VANETs." *Pervasive and Mobile Computing* 48 (2018): 43-58.

[25] Da Cunha, Felipe Domingos, Azzedine Boukerche, Leandro Villas, Aline Carneiro Viana, and Antonio AF Loureiro. "*Data communication in VANETs: a survey, challenges and applications.*" (2014).

[26] Schoch, Elmar, Frank Kargl, Michael Weber, and Tim Leinmuller. "Communication patterns in VANETs." *IEEE Communications Magazine* 46, no. 11 (2008): 119-125.

[27] Sichitiu, Mihail L., and Maria Kihl. "Inter-vehicle communication systems: a survey." *IEEE Communications Surveys & Tutorials* 10, no. 2 (2008): 88-105.

[28] I.V.T. Society, 1609.4-2010 – *IEEE standard for wireless access in vehicular environments (wave)–multi-channel operation (revision of IEEE STD 1609.4-2006).*

[29] DSRC, Dsrc, http://grouper.ieee.org/groups/scc32/dsrc/.

[30] Kenney, John B. "Dedicated short-range communications (DSRC) standards in the United States." *Proceedings of the IEEE* 99, no. 7 (2011): 1162-1182.

[31] Morgan, Yasser L. "Notes on DSRC & WAVE standards suite: Its architecture, design, and characteristics." *IEEE Communications Surveys & Tutorials* 12, no. 4 (2010): 504-518.

[32] Wang, Shie-Yuan, Chih-Che Lin, Kuang-Che Liu, and Wei-Jyun Hong. "On Multi-hop Forwarding over WBSS-based IEEE 802.11 (p)/1609 Networks." In *2009 IEEE 20th International Symposium on*

Personal, Indoor and Mobile Radio Communications, pp. 3040-3044. IEEE, 2009.

[33] Deng, Der-Jiunn, Hsin-Chin Chen, Han-Chieh Chao, and Yueh-Min Huang. "A collision alleviation scheme for IEEE 802.11 p VANETs." *Wireless Personal Communications* 56, no. 3 (2011): 371-383.

[34] Kumar, M. Abhilash, and A. Rama Mohan Reddy. "A Survey on Vehicular Ad Hoc Networks Routing Protocols." *i-manager's Journal on Mobile Applications and Technologies* 3, no. 2 (2016): 19.

[35] Dutta, Rajashree, and Ranjana Thalore. "A Review of Various Routing Protocols in VANET." *International Journal of Advanced Engineering Research and Science* 4, no. 4 (2017).

[36] Perkins, Charles E., and Pravin Bhagwat. "Highly dynamic destination-sequenced distance-vector routing (DSDV) for mobile computers." In *ACM SIGCOMM computer communication review*, vol. 24, no. 4, pp. 234-244. ACM, 1994.

[37] Clausen, Thomas, Philippe Jacquet, Cédric Adjih, Anis Laouiti, Pascale Minet, Paul Muhlethaler, Amir Qayyum, and Laurent Viennot. "*Optimized link state routing protocol (OLSR).*" (2003).

[38] Pei, Guangyu, Mario Gerla, and Tsu-Wei Chen. "Fisheye state routing: A routing scheme for ad hoc wireless networks." In *2000 IEEE International Conference on Communications. ICC 2000. Global Convergence through Communications. Conference Record*, vol. 1, pp. 70-74. IEEE, 2000.

[39] Perkins, Charles E., and Elizabeth M. Royer. "Ad-hoc on-demand distance vector routing." In *Proceedings WMCSA'99. Second IEEE Workshop on Mobile Computing Systems and Applications*, pp. 90-100. IEEE, 1999.

[40] Johnson, David, Yin-chun Hu, and David Maltz. *The dynamic source routing protocol (DSR) for mobile ad hoc networks for IPv4*. Vol. 260. RFC 4728, 2007.

[41] Park, V., and S. Corson. "Temporally-ordered routing algorithm (tora) version 1." *Functional specification IETF Internet draft* (2001).

[42] Haas, Zygmunt J., and Marc R. Pearlman. "The zone routing protocol (ZRP) for ad hoc networks (Internet-Draft)." *Mobile Ad-hoc Network (MANET) Working Group, IETF* (1998).

[43] Nikaein, Navid, Christian Bonnet, and Neda Nikaein. "Harp-hybrid ad hoc routing protocol." In *Proceedings of international symposium on telecommunications (IST)*, pp. 56-67. 2001.

[44] Luo, Yuyi, Wei Zhang, and Yangqing Hu. "A new cluster based routing protocol for VANET." In *2010 second international conference on networks security, wireless communications and trusted computing*, vol. 1, pp. 176-180. IEEE, 2010.

[45] Santos, R. A., R. M. Edwards, and A. Edwards. "Cluster-based location routing algorithm for vehicle to vehicle communication." In *Proceedings. 2004 IEEE Radio and Wireless Conference (IEEE Cat. No. 04TH8746)*, pp. 39-42. IEEE, 2004.

[46] Xia, Yang, Chai Kiat Yeo, and Bu Sung Lee. "Hierarchical cluster based routing for highly mobile heterogeneous MANET." In *2009 International Conference on Network and Service Security*, pp. 1-6. IEEE, 2009.

[47] Ibrahim, Khaled, Michele C. Weigle, and Mahmoud Abuelela. "p-IVG: Probabilistic inter-vehicle geocast for dense vehicular networks." In *VTC Spring 2009-IEEE 69th Vehicular Technology Conference*, pp. 1-5. IEEE, 2009.

[48] Joshi, Harshvardhan P. "*Distributed robust geocast: A multicast protocol for inter-vehicle communication.*" (2007).

[49] Rahbar, Hamidreza, Kshirasagar Naik, and Amiya Nayak. "DTSG: Dynamic time-stable geocast routing in vehicular ad hoc networks." In *2010 The 9th IFIP Annual Mediterranean Ad Hoc Networking Workshop (Med-Hoc-Net)*, pp. 1-7. IEEE, 2010.

[50] Durresi, Mimoza, Arjan Durresi, and Leonard Barolli. "Emergency broadcast protocol for inter-vehicle communications." In *11th*

International Conference on Parallel and Distributed Systems (ICPADS'05), vol. 2, pp. 402-406. IEEE, 2005.

[51] Korkmaz, Gökhan, Eylem Ekici, Füsun Özgüner, and Ümit Özgüner. "Urban multi-hop broadcast protocol for inter-vehicle communication systems." In *Proceedings of the 1st ACM international workshop on Vehicular ad hoc networks*, pp. 76-85. ACM, 2004.

[52] Nagaraj, Uma, and Poonam Dhamal. "Broadcasting routing protocols in VANET." *Network and Complex Systems* 1, no. 2 (2012): 13-19.

[53] Karp, Brad, and Hsiang-Tsung Kung. "GPSR: Greedy perimeter stateless routing for wireless networks." In *Proceedings of the 6th annual international conference on Mobile computing and networking*, pp. 243-254. ACM, 2000.

[54] Lochert, Christian, Hannes Hartenstein, Jing Tian, Holger Fussler, Dagmar Hermann, and Martin Mauve. "A routing strategy for vehicular ad hoc networks in city environments." In *IEEE IV2003 Intelligent Vehicles Symposium. Proceedings (Cat. No. 03TH8683)*, pp. 156-161. IEEE, 2003.

[55] Seet, Boon-Chong, Genping Liu, Bu-Sung Lee, Chuan-Heng Foh, Kai-Juan Wong, and Keok-Kee Lee. "A-STAR: A mobile ad hoc routing strategy for metropolis vehicular communications." In *International Conference on Research in Networking*, pp. 989-999. Springer, Berlin, Heidelberg, 2004.

[56] Lochert, Christian, Martin Mauve, Holger Füßler, and Hannes Hartenstein. "Geographic routing in city scenarios." *ACM SIGMOBILE mobile computing and communications review* 9, no. 1 (2005): 69-72.

[57] Naumov, Valery, and Thomas R. Gross. "Connectivity-aware routing (CAR) in vehicular ad-hoc networks." In *IEEE INFOCOM 2007-26th IEEE International Conference on Computer Communications*, pp. 1919-1927. IEEE, 2007.

[58] Rondinone, Michele, and Javier Gozalvez. "Contention-based forwarding with multi-hop connectivity awareness in vehicular ad-hoc networks." *Computer Networks* 57, no. 8 (2013): 1821-1837.

[59] Lee, Kevin C., Uichin Lee, and Mario Gerla. "TO-GO: TOpology-assist geo-opportunistic routing in urban vehicular grids." In *2009 Sixth International Conference on Wireless On-Demand Network Systems and Services*, pp. 11-18. IEEE, 2009.

[60] Zhao, Jing, and Guohong Cao. "VADD: Vehicle-assisted data delivery in vehicular ad hoc networks." *IEEE transactions on vehicular technology* 57, no. 3 (2008): 1910-1922.

[61] Wang, Xin, Changle Li, Lina Zhu, and Chunchun Zhao. "An effective routing protocol for intermittently connected vehicular ad hoc networks." In *2013 IEEE Wireless Communications and Networking Conference (WCNC)*, pp. 1750-1755. IEEE, 2013.

[62] Wang, Xin, Changle Li, Lina Zhu, and Chunchun Zhao. "An effective routing protocol for intermittently connected vehicular ad hoc networks." In *2013 IEEE Wireless Communications and Networking Conference (WCNC)*, pp. 1750-1755. IEEE, 2013.

[63] Shafiee, Kaveh, and Victor CM Leung. "Connectivity-aware minimum-delay geographic routing with vehicle tracking in VANETs." *Ad Hoc Networks* 9, no. 2 (2011): 131-141.

In: Vehicular Ad-Hoc Networks ISBN: 978-1-53618-038-1
Editors: G. G. Md. Nawaz Ali et al. © 2020 Nova Science Publishers, Inc.

Chapter 2

ELLIPTIC CURVE CRYPTOGRAPHY IN INTELLIGENT TRANSPORT SYSTEMS

Muhammad Faisal Siddiqui, Muhammad Awais Javed [*]
and Rida Fatima
Department of Electrical and Computer Engineering, COMSATS
University Islamabad, Pakistan

ABSTRACT

The development of smart transportation systems all over the world is providing a lot of benefits and potential applications. Intelligent transport systems (ITS) aim to provide a safer, reliable and smarter transportation system with the help of wireless communications. Security and privacy are the major concerns when it comes to Intelligent Transport System, where the entire network is communicating wirelessly. Multiple cryptographic algorithms could be used to provide safe and reliable communication in ITS, however, the selected cryptographic algorithm should provide low latency. Elliptic Curve Cryptography (ECC) is a well-known technique to secure data from various tampering attacks. ECC uses Elliptic Curve Digital Signature Algorithm (ECDSA) for message

* Corresponding Author's Email: awais.javed@comsats.edu.pk.

signature and verification. The verification of messages containing ECDSA signatures is a lengthy and time-consuming process. ECDSA verification requires double scalar multiplication on the Elliptic curve points. For real time systems, the verification of such messages becomes a challenging task. To speed up the entire verification process, we focus on improving the double scalar multiplication time. In this chapter, endomorphism is exploited which divides the scalars into equal bit length parts. This reduces the range of data and parallel processing can be utilized efficiently. Reducing the range of data needs to be computed also reduces the memory utilization of the hardware; as the generated lookup tables are smaller in size. Furthermore, parallel processing provides faster computations. The proposed architecture is not fixed for a specific public key, thus making the system generic for any public key Q. The lookup tables are generated in real-time, thus providing better throughput. Results show that using multilevel endomorphism improves the total latency of the system by 2.3 times.

Keywords: intelligent transport systems; elliptic curve cryptography; FPGA

INTRODUCTION

The Internet of Things (IoT) is an enabling technology that allows devices such as smartphones and sensors to communicate with the Internet. IoT plays a vital role in the advanced age of intelligent networks [1]. One of the emerging examples of IoT is the Intelligent Transportation System (ITS) where the entire road network can communicate with each other using sensors and other wireless communication modules [2].

ITS provides many applications for traffic management and safe road driving. By connecting vehicles wirelessly, the vision of each vehicle increases beyond the line of sight. The shared information can be used by vehicles to make intelligent driving decisions. In addition, smart city data centers can collect this information to manage the traffic efficiently and develop a congestion free road network.

A key requirement of ITS is fast and reliable data sharing. This means that vehicles should exchange data with each other in a secure manner and avoid various possible security attacks such as man-in-the-middle, jamming, eavesdropping, denial of service, etc. A robust cryptographic algorithm is essential to ensure the reliability of the data. In addition, various types of data integrity attacks should also be mitigated.

Cryptographic algorithms can improve the security level of the information exchange; however, they could result in high end-to-end delay and increased latency of the shared data. This can be a big hindrance in intelligent and quick decision making by the vehicles and negatively impact the safety applications. Thus, the cryptographic algorithms in ITS must be fast and use low computational resources [3].

Elliptic Curve Cryptography (ECC) is the most widely used public-key cryptographic algorithm for ITS as it provides robust security with a low message overhead [5]. Thousands of messages are broadcasted per second by the vehicles and each of these messages contains the ECDSA signature for security purposes. At the receiver, verification of these messages may overload the hardware due to the high computation cost involved. Thus, to efficiently verify such messages, a fast processing element is required that can provide low latency and high throughput.

Field Programmable Gate Arrays (FPGAs) provides a platform for fast and real-time implementation of security protocols [6]. As double scalar multiplication is the major delay causing agent in ECDSA verification procedure, optimized multiplication techniques are needed to improve the overall system's performance. The significance of speed in signature verification of ECDSA creates the necessity to make application-specific hardware for a real-time solution of ECC. The double scalar multiplication consumes the maximum computational time in the overall signature verification process in ECDSA. This affects the overall throughput of the system. Due to this computational delay, ECDSA implementation for real-time systems is a challenging task. FPGAs thus offers a higher computational speed as compared to any software-based implementation on a multi-core CPU.

We propose a method to reduce the verification time of ECDSA signature by improving the computational time of double scalar multiplication, using multilevel endomorphism and pre-computational method. The main contributions of the proposed work in this chapter are as follows:

- an architecture to reduce the verification time of digital signatures with the help of multi-level endomorphism.
- reduces the data manipulation cost by pre-computation of the elliptic curve multiple points used for verification purposes.
- develops an optimized real-time solution for the ECC using FPGA.

We start the chapter by introducing ITS and its applications. This is followed by an overview of security in ITS and the ECDSA algorithm. We explain the proposed architecture for the fast verification of ECDSA signatures along with the results. Finally, conclusions are provided, and future work is presented.

ITS Applications and Security

The concept of smart cities (revamping the cities into digital societies) with every facet making the life of citizens very easy, the Intelligent Transport System becomes the crucial component amongst all other viable components. ITS aims to achieve traffic efficiency by minimizing traffic problems, facilitating citizens' safe mobility for their daily routines by providing them prior information about traffic congestion and emergency situations, real-time availability of transport, seat availability information, etc. This not only reduces the travel time of the passengers but also provides them with safety and comfort [11]. As it provides endless possibilities, ITS has now become a wide field of interest for the researchers and many organizations of the world to use ITS for providing solutions to their road problems [12].

ITS applications range from safety, traffic management to infotainment and comfort. Few of the key ITS applications include safe lane change, optimal traffic signal control, emergency vehicle notification system, smart car parking system, etc.

Security in ITS

Privacy and security are the primary concerns when it comes to vehicular communication. The Intelligent Transport System (ITS) establishes wireless communication between vehicles and the fixed infrastructure, thus allowing vehicles to be aware of neighborhood traffic conditions. To ensure reliable communication in ITS, it is necessary to protect the system from the tampering attacks and other security threats, where some malicious users may insert false information to corrupt the data. To address this concern, an efficient cryptographic technique is adopted by the ITS standards called Elliptic Curve Cryptography (ECC) [3]. Elliptic Curve Cryptography (ECC) is a well-known public key cryptographic algorithm that provides robust security with a small message overhead.

Elliptic Curve Cryptography

The use of public-key cryptographic algorithms has blossomed over the past few decades. A lot of methods of encryption, authentication, signing and key exchange have been developed in the past few years [7]. One of the most known public-key cryptographic algorithms is Elliptic Curve Cryptography (ECC).

Elliptic curve cryptography has been widely used as an alternative approach for implementing public-key cryptography. Public key algorithms allow the sharing of keys and information among multiple participants in a complex information system. The elliptic curve has evolved as a broad field for the Public Key Cryptography (PKC) systems. Separate keys are used to encode and decode the data in the (PKC) system [9]. The keys are divided as:

- Public key
- Private key

The security strength of the cryptographic protocols depends on the size of the key, larger the key size more secure the encoded data is. This results in mathematical prime factorization and discrete logarithmic problems. Elliptic curve cryptography was presented as a solution to the previous (PKC) system, providing the same level of security for the encoded data but with a smaller key size. The security in Elliptic Curve Cryptography lies in the difficulty to solve the discrete logarithm. ECC provides high levels of security with very small key sizes as compared to other cryptosystems such as Rivest–Shamir-Adleman (RSA). ECC has become the faster mode of cryptography being used for multiple cryptographic operations like digital signatures, secret key sharing, and message encryption. The Table 1 below briefly describes the key size of ECC and RSA, the key generation time, the signature generation time and the signature verification time.

Table 1. A comparison of ECC and RSA [10]

Security Bits	**Key sizes (in bits)**		**Key generation time (s)**		**Signature generation time (s)**		**Signature verification time (s)**	
	ECC	RSA	ECC	RSA	ECC	RSA	ECC	RSA
80	163	1024	0.08	0.16	0.15	0.01	0.23	0.01
112	233	2240	0.18	7.47	0.34	0.15	0.51	0.01
128	283	3072	0.27	9.80	0.59	0.21	0.86	0.01
192	409	7680	0.64	133.9	1.18	1.53	1.80	0.01
256	571	15360	1.44	679.1	3.07	9.20	4.53	0.03

It can be seen from Table 1 that the key sizes of ECC is very small as compared to the RSA algorithm, despite the small key size ECC can provide the same level of security. This gives ECC edge over other well-known cryptographic algorithms.

Elliptic Curve

The generalized definition of an elliptic curve E over a field K is mathematically given as:

$$E: y^2 + a_1xy + a_3 \tag{1}$$

This equation is known as the Weierstrass equation, where

$$a_1, a_2, a_3 \ldots . \in k$$

Domain Parameters of the Elliptic Curve

The global domain parameters used in elliptic curve cryptographic implementations are a, b, G, n, q. "**a**" and "**b**" are the points defining the elliptic curve, G is the base point or the point generator, $(x_{G,}y_G)$ is the point selected for the algorithm implementation and "n" is the range of integers from $(1, n - 1)$.

Point Addition and Point Multiplication

For any point P and Q on the elliptic curve, the point addition is defined as "if we draw a line from point P to point Q, this line will intersect at a point R the reflection of the point R on the x-axis is the addition (P+Q) of point P and Q" [9].

The point doubling is 2P where P=Q, so if we draw a tangent line from P and it cuts the elliptic curve at some point (–R), the reflection of this point on the x-axis is called the point doubling. The point multiplication is performed by repeated point additions and point doublings.

$$K = Q * P \tag{2}$$

Different Multiplication Techniques

A lot of work has been done previously to improve the verification time of ECDSA in ITS. Many optimized solutions have been presented to simplify the most computationally complex part of ECDSA verification that is double scalar multiplication.

In [5], Peter Rombouts introduces a novel approach by combining the basic finite field multiplication with the simple arithmetic unit (SAU), which contributes to achieving high overall performance. Top-down and bottom-up techniques are introduced. Different methods of multiplication are proposed to achieve a high throughput system on the cost of increased hardware. The basic technique is that by using different word sizes, in which a finite- field multiplier is constructed using three different architectures namely; multiply and reduce, interleaved and karatsuba.

In [13], a double scalar multiplication technique has been introduced. Double scalar multiplication is of the form, where k and l are two scalars and G is a fixed point and Q is any arbitrary point. The techniques used are two single scalar multiplications; where there are two separate single scalar multiplications and then finally add the results of both scalar multiplications. Thus, it reduces the point additions to half due to the pre-computation of the scalars but inversely increases the points twice. Next is the interleaving method for double scalar multiplication; where first both the points are added and then scalars are scanned simultaneously. This method also reduces the point doubling yet introduces the delay in the system due to the pre-computation.

Another method investigated is a joint sparse form [14], in this technique, the weight of the binary numbers is reduced using signed binary representations for the scalars k and l, and are pre-computed. Using this method along with the interleaving method reduces the point multiplications but doubles the hardware.

Another very efficient technique implemented is a common subexpression technique, that uses the matrix multiplication and reduces the computational time using parallel processing techniques [15].

Verdict based multiplier architecture [16] is one of the optimized architectures that consumes less computation time and power. The verdict multiplier combines the floating-point multipliers with the signed multipliers. Hence reduces the complexity and cost of the system.

A book [17], the author presents the work where the technique used is Partial Parallel Shifting Multiplier (PPSM) and parallel complementary method. Parallel processing and the spat architecture highly improve the overall performance of the system along with the hardware design as it uses fewer logic gates and LUTs.

A redundant basis multiplication for the ECC verifications has been proposed by Pramod Kumar Mehar [18], a novel reduction technique has been discussed to obtain high throughput by using digital serial implementation.

D. Al-Khalili and N. Chabini [19] proposed an asymmetric, un-pipelined; signed and unsigned multipliers blocks. Partial products are organized in various configurations and the additions of the products are realized in an optimized manner. This approach reduces the critical path delay and the FPGA resource utilization is also reduced, thus improving the throughput of the system and reducing the hardware delay, hence improving the overall system performance.

In the latest research [13], the endomorphism is introduced for double scalr multipliation. The scalar having the range [1, n-1] is subdivided into equal bit length scalars having the range half as compared with the desired scalar. The range is reduced thus the computation is also reduced because of parallel processing. The extra memory blocks are used for pre-calculating the possible scalar points, which became the bottleneck of this technique. Another drawback of this technique is that the architecture is fixed for a specific public key if the public key is changed the entire architecture must be redesigned.

The techniques mentioned above provide high throughput, at the cost of large multiplier sizesmeans extra DSP blocks or/and large memory usage. The main motivation of this research is to find optimized ways to reduce the size of the double scalar multiplierand reducesthe computational time as well. Furthermore, the optimized ECC multiplier would be implemented on hardware to improve the latency of the overall system, thus making it applicable for ITS applications. The critical analysis of the different methods proposed in the literature review for achieving high throughput is summarized in the table given below:

Table 2. Comparison of different multiplication techniques

Serial no.	Technique	Reference	Pros	Cons
1	Basic finite field multiplication with the simple arithmetic unit (SAU)	[5]	Low latency	Increased hardware
2	Two single scalar multiplications	[3]	Reduce the point additions by pre-computation	Increases the points to double
3	Interleaved	[3]	Reduces the number of point doublings	------------
4	Verdict based multiplier	[8]	Reduces complexity, cost and power	------------
5	Partial parallel shift multiplier	[9]	Low cost and throughput	Extensive hardware
6	Asymmetric pipelined signed and unsigned multiplier	[11]	Minimization of critical path	-------------
7	Redundant base multiplication	[10]	Negligible hardware cost	High critical path delay
8	Joint sparse	[6]	Speedup as pre-computations are involved	requires subtractions for negative digits
9	Common sub-expression technique	[7]	Low cost and hardware	----------------
10	Efficiently computable endomorphism	[13]	Speedup due to parallel processing and scalar division.	Extra memory blocks and fixed architecture

PROPOSED ARCHITECTURE

Using ECDSA, messages along with their signature are received at the vehicle receivers. Once the message is received at the receiving side, it is verified and decrypted. The messages containing the ECDSA signatures take time for verification as the double scalar multiplication consumes most of the verification time chunk.

The double scalar multiplication is of the form:

$$X = U_1.G + U_2.Q$$

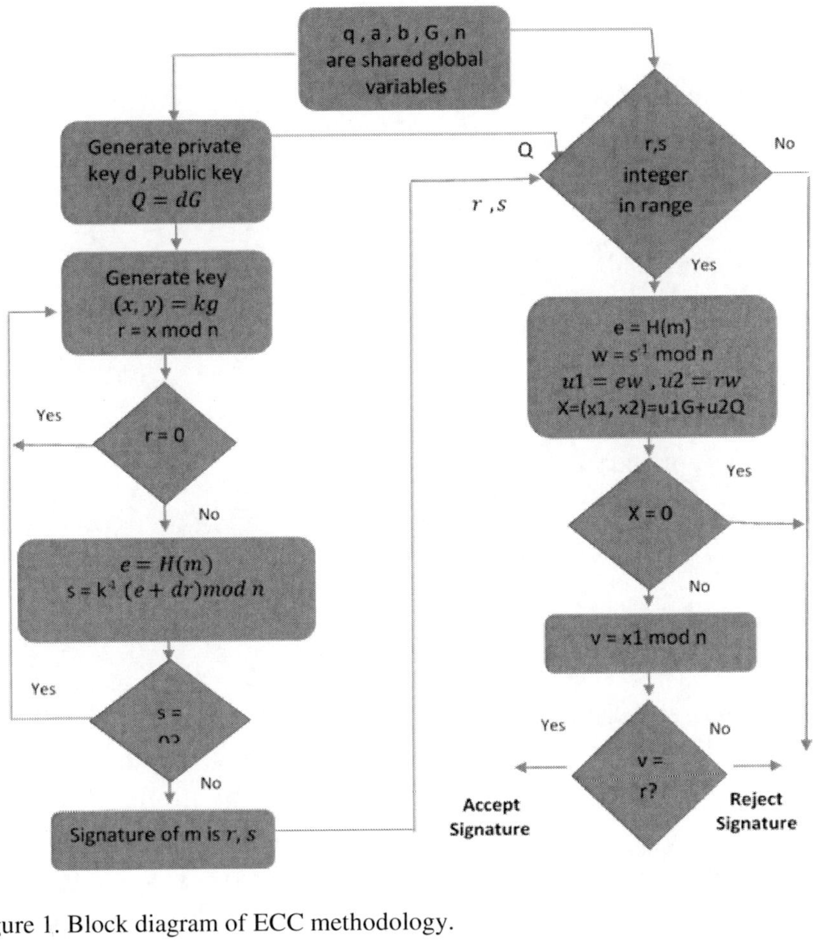

Figure 1. Block diagram of ECC methodology.

A block diagram depicting the basic ECC Algorithm showing the receiver side and sending side is given in Figure 1. Introducing a multilevel endomorphism for computing the double scalar multiplication will reduce the latency of the verification process. The following section explains the utilization of endomorphism for improving the double scalar multiplication time.

Endomorphism

The term endomorphism is derived from a Greek adverb endow (inside) morphosis ("to form"," to shape").

Let X be a set, F be a sigma-algebra on X, and m is a probability measure. A map T:X→X is called endomorphism if the following conditions are met [20]:

1: T is surjective

2: T is measurable

3: $m(T^{-1}A) = m(A)$ where $A \epsilon$F.

φ is the endomorphism acting on the elliptic curve E defined over a finite field F_Q. φ acts on the subgroup of elliptic curve E as λ. Lambda is defined as the characteristic root of $\varphi \, mod \, n$. φ is exploited by decomposing the scalar into equal bit length parts. Thus, the range of each part is reduced to half.

Efficient Usage of Endomorphism in ECDSA

Despite the small key sizes of ECDSA, it is usually more computationally extensive at the receiver side. The verification of ECDSA requires a double scalar multiplication of the form $R = k.G + l.Q$, where G is the base point on the elliptic curve, Q is the public key or the arbitrary point on the curve, "k" and "l" are the integers ranging from [1, m-1], where m is the order of base point and must be a prime number. Usually, R is computed in a simultaneous way that is with the joint point doubling, so most of the point doublings are calculated at a time [21]. To speed up the multiplication process the integers are split into different number of parts.

The integer $k = k_1 + k_2 + k_3$ is split into two integers of equal size [8]. And the scalar multiplication is computed as:

$$k_1 = k_{121} + \lambda_1 k_{122} + \lambda_2 k_{12} \quad (3)$$

The real-world benefits of this technique are multiple implementation options and the tradeoffs between execution time and hardware area. Our design allows the user to fine-tune the implementation according to demand in hand. In this research, we are using a prime field. First, we generate the point multiplier and point adder then a lookup table for storing values from point P (Px, Py) to 8P $(8Px, 8Py)$. In [13] the prime field is 257, whereas in our system the security requirements are fulfilled by using a prime field of the 237-bit field providing enough security for ITS based applications as discussed in the literature review [13]. By using endomorphism, we have reduced the computation time as compared to [13]. The range is from (P, 256P), using endomorphism we reduce our range to (P, 8P). As we know that double scalar multiplication is of the form:

$$X = U_{1.}G + U_2 . Q \quad (4)$$

Let $k_1 = U_1;$

The range is from [1, p-1]

Applying endomorphism

$$k_1 = k_{11} + \lambda_2 k_{12} \quad (5)$$

Now the range is reduced to:

$$[1 - \sqrt{p} - 1]$$

Further reducing the integer k_1

$$k_{11} = k_{121} + \lambda_1 k_{122} \tag{6}$$

So, the equation becomes

$$k_1 = k_{121} + \lambda_1 k_{122} + \lambda_2 k_{12} \tag{7}$$

Thus by applying endomorphism the range of the integers k_{121}, k_{122}, k_{12} becomes half of k_1. The major advantage is the lesser computation because the individual calculations of each sub divided part is much lesser than the computation time required to calculate the overall scalar multiplication.

Calculations Using Endomorphism

Considering an example where the prime number $p \leq 23$, the range for $p \leq 23$ can be defined as [1, n-1]. By applying endomorphism, the range reduces and becomes;

$$(1, \sqrt{p} - 1) = (1, \sqrt{23} - 1) = (1P - 4P)$$

By taking advantage of the endomorphism range that was first [1, p-1], is now reduced to half. Thus, we only need four points to reach any point until 23P. For bigger architectures, the advantages of endomorphism become more prominent. For example, if we take prime number $p \leq 237$, only eight (1P - 8P) points are required to reach any point till 237. The use of endomorphism becomes prominent for bigger architectures. Only two levels are required to reach 23P, and by increasing only 1 more level, any point till 237P can be achieved. The block diagram of the architecture is given in Figure 2.

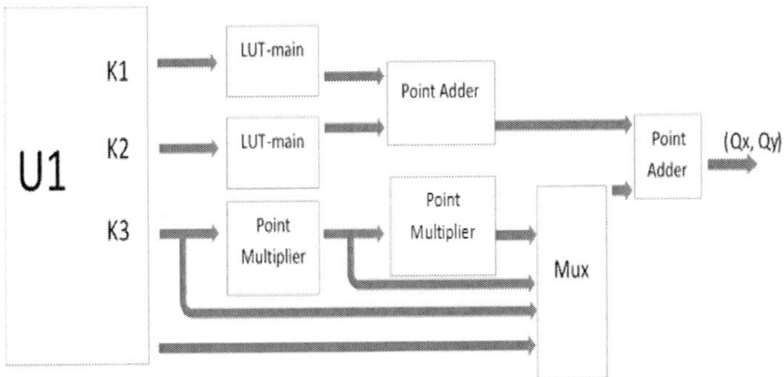

Figure 2. Block diagram of the proposed architecture.

Point Adder

For any point P and Q on the elliptic curve, the point addition is defined as "if we draw a line from point P to point Q, this line will intersect at a point R the reflection of the point R on the x-axis is the addition (P+Q) of point P and Q [9]. The adder is used to add two different points on the curve P (Px, Py) and Q (Qx, Qy) and providing the resultant $R = P + Q$;$R_x = P_x + Q_x$; $R_y = P_y + Q_y$.The resultant X_R and Y_R are calculated using the following formula:

$$\lambda = \frac{Y_Q - Y_P}{X_Q - X_P} \tag{8}$$

$$X_R = (\Delta^2 - X_P - X_Q) mod p \tag{9}$$

$$Y_R = (\Delta * (X_p - X_R) - Y_P) mod p \quad (10)$$

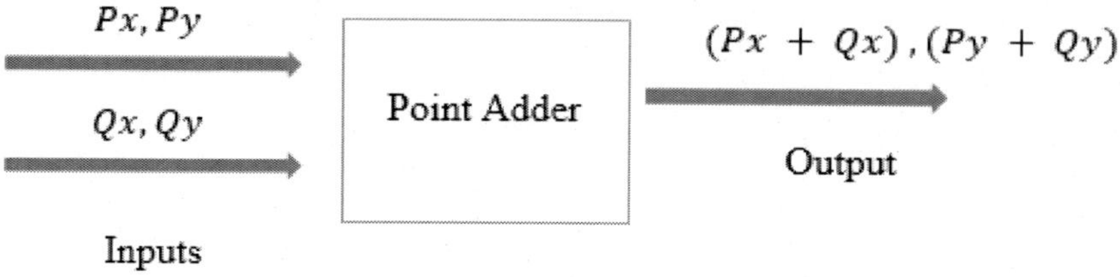

Figure 3. Block diagram of point adder.

$$\lambda = \frac{Y_Q - Y_P}{X_Q - X_P}$$

The points Xp, Yp, Xq, Yq are the points on the curve and serve as input to this block.

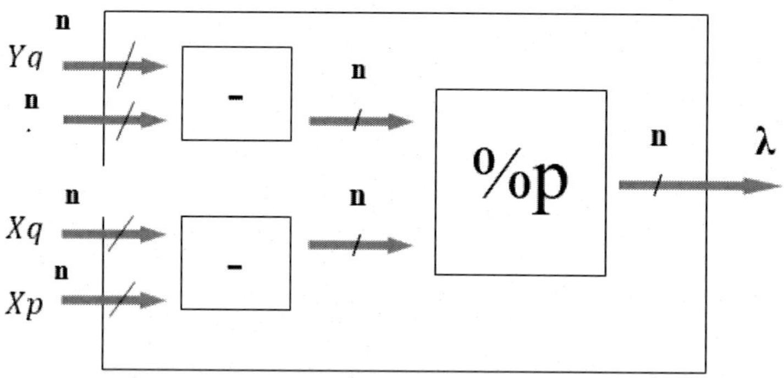

Figure 4. Proposed architectural diagram to calculate λ.

$$X_R = (\lambda^2 - X_P - X_Q) mod p$$

Xr is the resultant, whereas Xp and Xq are the inputs.

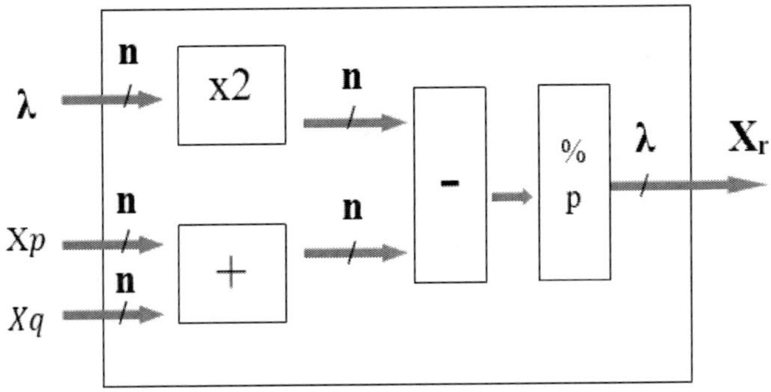

Figure 5. Proposed architectural diagram to calculate resultant ***Xr***.

$$Y_R = (\Delta * (X_p - X_R) - Y_P) mod p$$

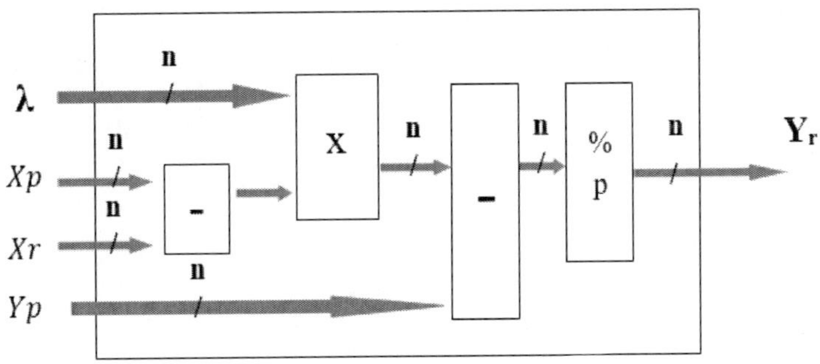

Figure 6. Proposed architectural diagram to calculate resultant ***Yr***.

Point Multiplier

The multiplier is 2P multiplier providing us with the point P multiplied by 2 at the result. The point multiplier is adding a point twice which is not possible by point adder, this is called point doubling. For multiplication, we first find the following parameters:

$$\lambda = \left(\frac{3x^2+a}{2y_p}\right) mod p \tag{11}$$

$$X_R = (\Delta^2 - 2x_p)modp \tag{12}$$

$$Y_R = \left(\Delta * \left(x_p - x_R\right) - y_p\right)modp \tag{13}$$

The total computational time of the propsed point multiplier module is 151.26ns.

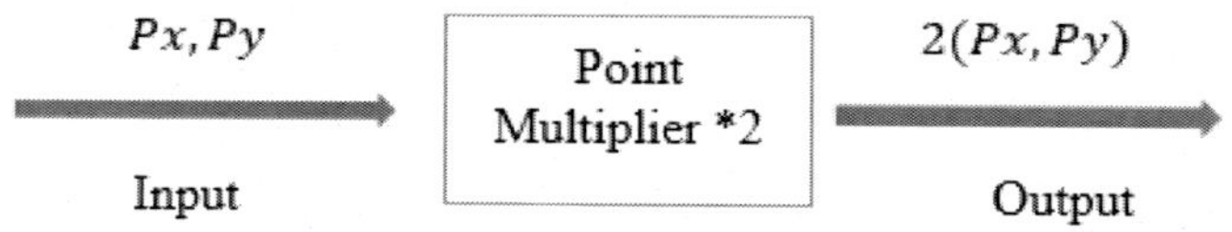

Figure 7. Block diagram of the proposed point multiplier.

Internal architectural diagrams of point multiplier:

$$\lambda = \left(\frac{3x^2 + a}{2y_p}\right)modp$$

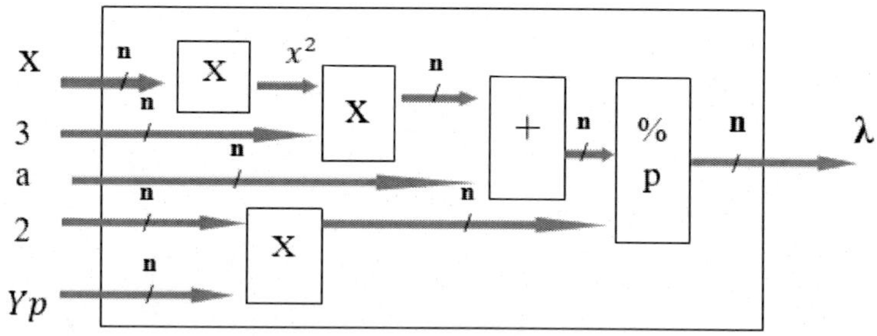

Figure 8. Proposed architectural diagram to calculate λ.

$$X_R = (\lambda^2 - 2x_p)modp$$

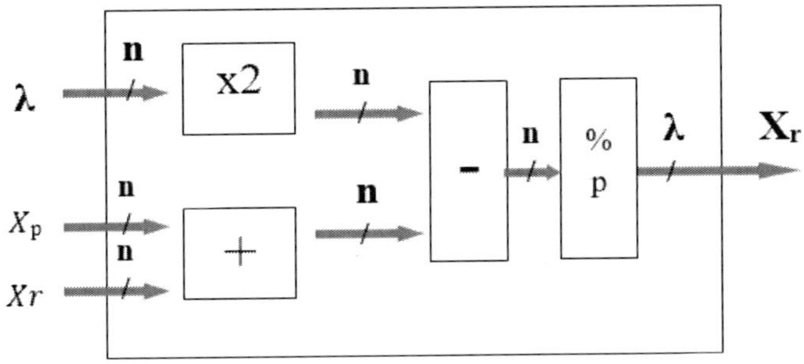

Figure 9. Proposed architectural diagram to calculate resultant X.

$$Y_R = (\lambda * (x_p - x_R) - y_p) mod p$$

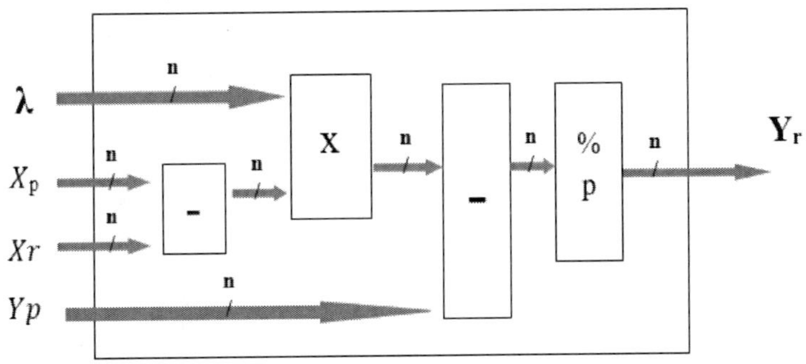

Figure 10. Proposed architectural diagram to calculate resultant $\boldsymbol{Y_r}$.

LUT Main

By applying the endomorphism, the calculations are reduced to squared root of the initial prime number field. By the choice of the prime number p the lookup table is generated. In this research two different cases are discussed:

1. $p \leq 23$
2. $p \leq 237$

When $p \leq 23$ we only need to go to 2 levels and the proposed architecture becomes less computationally extensive. For $p \leq 237$only 3 levels are required for the complete calculations.

Proposed Architecture for $p \leq 23$

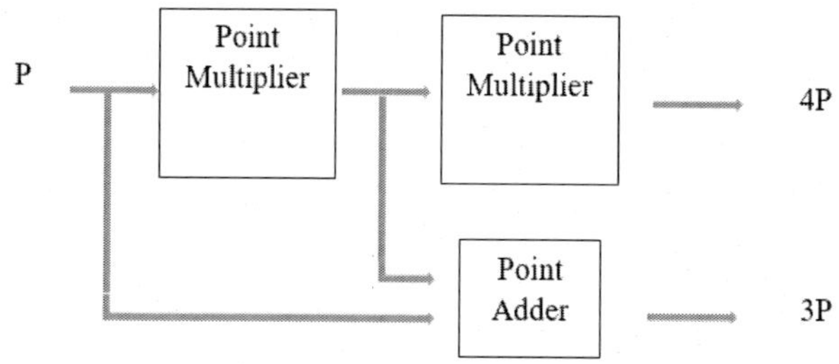

Figure 11. Block diagram of the proposed architecture for$\boldsymbol{p \leq 23}$.

Total Delay

Case 1: $p \leq 23$

For $p \leq 23$ we only need 2 multipliers and 1 adder, so the total latency is the delay caused by 2-point multipliers.

Latency: 2× (point multiplier)

Case 2: $p \leq 237$

When $p \leq 237$, the level of calculations goes to level 3, and the computational time slightly increases.

Latency: 3× (point multiplier).

Architecture for $p \leq 237$

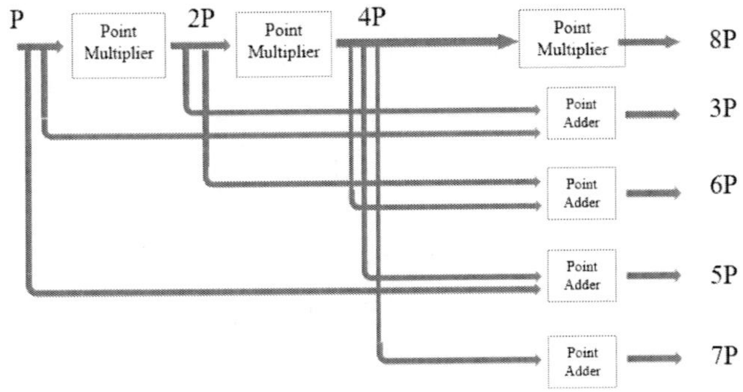

Figure 12. Proposed block diagram for $p \leq 237$.

We only need to go to 8P as the range of prime is $p \leq 237$. The lookup table stores points till 8P.

P
2P
3P
4P
5P
6P
7P
8P

Figure 13. Block diagram of the LUT.

A total of 4 adders are used in the LUT module and 3 multipliers are taken to utilization. The first point on the curve P is taken as input, passed through multiplier and a 2P point is obtained at the output of the point multiplier, similarly again passing from the point multiplier the point is added to itself and 4P is generated at the output. The same process is repeated to duplicate any point on the elliptic curve. For the proposed architecture a point doubling till 8P is required.

For point addition, point doublings are bought to utilization. For example, to generate a point 3P, a point P is added to the point 2P. The same goes for any point needed to be generated.

U1-k Module

This module is designed in a way to select between multiple values of points k. For the proposed architecture only four points are required as prime number 23 is taken as an example. The maximum value of 'k' is only needed to go at maximum value of U1 i.e., 22.

$$p \leq 23 \rightarrow \sqrt{p} - 1$$

$$\sqrt{23} - 1 = 4.7 - 1$$

$$3.71 \cong 4$$

Another example of point taken is p = 29

$$p >= 29 \sqrt{p} - 1$$

$$\sqrt{29} - 1 = 5.7 - 1$$

$$4.73 \cong 5$$

So whenever $23 < p \leq 237$, the architectural levels move from level 2 to level 3. Effecting the latency of the system but the throughput of the system will remain significantly same for 2 level architecture.

The range for k1, k2, k3 is [0- 4].

In the proposed module the input is taken from the value of U1[1-22] and correspondingly the values of k1, k2, k3 and lambda are generated.

$$U1 = K_1 + \lambda_1 K_2 + \lambda_2 K_3$$

The different values of U1 ranging from [1-22] are given in the following table:

Table 3. For p≤ 23; the values of $k1, k2, k3, \lambda 1, \lambda 2$

U1	k_1	k_2	λ_1	k_3	λ_2
1	1	0	0	0	0
2	2	0	0	0	0
3	3	0	0	0	0
4	4	0	0	0	0
5	4	1	1	0	0
6	4	2	1	0	0
7	4	3	1	0	0
8	4	4	1	0	0
9	4	4	1	1	1
10	4	4	1	2	1
11	4	4	1	3	1
12	4	4	1	4	1
13	4	1	1	4	2
14	4	2	1	4	2
15	4	3	1	4	2
16	4	4	1	4	2
17	1	0	1	4	4
18	2	0	1	4	4
19	3	0	1	4	4
20	4	0	1	4	4
21	4	1	1	4	4
22	4	2	1	4	4

RESULTS

As a solution to the double scalar problem in ECDSA signature verification, our research has proposed an architecture based on multilevel endomorphism that has reduced the computational complexity, improved the double scalar multiplication time and improved the memory utilization. The results are classified as validated results and computational results.

For validation purposes elliptic curve over the Zp, choosed is given below:

$$y^2 modp = (x^3 + ax + b)modp \quad (14)$$

Validation Results

For $p \leq 23$:

Table 4(a). Validated results for $p \leq 23$

P	P	A	b	Actual			Calculated		
				2p	3p	4p	2p	3p	4p
23	(0,1)	1	1	(6,19)	(3,13)	(13,16)	(6,19)	(3,13)	(13,16)
23	(6,4)	1	1	(13,7)	(7,12)	(5,4)	(13,7)	(7,12)	(5,4)
23	(12,19)	1	1	(5,19)	(6,4)	(17,3)	(5,19)	(6,4)	(17,3)
23	(0,22)	1	1	(6,4)	(3,10)	(13,7)	(6,4)	(3,10)	(13,7)

For $p \leq 23$, the total latency is of 2 multipliers. Here it is observed that our overall computational time has been reduced.

For $p \leq 237$:

Table 4(b). Validated results for$p \leq 237$

p	P	A	B	Result Type	2p	3p	4p	5p	6p	7p	8p
29	(0,1)	1	1	Actual	(6,19)	(3,13)	(13,16)	(18,3)	(7,11)	(11,3)	(5,19)
				Calculated	(6,19)	(3,13)	(13,16)	(18,3)	(7,11)	(11,3)	(5,19)
29	(6,4)	1	1	Actual	(13,7)	(7,12)	(5,4)	(12,19)	(17,3)	(4,0)	(17,20)
				Calculated	(13,7)	(7,12)	(5,4)	(12,19)	(17,3)	(4,0)	(17,20)
29	(12,19)	1	1	Actual	(5,19)	(6,4)	(17,3)	(9,19)	(10,9)	(12,11)	(4,6)
				Calculated	(5,19)	(6,4)	(17,3)	(9,19)	(10,9)	(12,11)	(4,6)
29	(0,22)	1	1	Actual	(6,4)	(3,10)	(13,7)	(18,20)	(7,12)	(11,20)	(5,4)
				Calculated	(6,4)	(3,10)	(13,7)	(18,20)	(7,12)	(11,20)	(5,4)

For $p \leq 237$ the total latency is of 4 multipliers respectively, the computational time of point multiplier is more than point adder, so in this case our overall computational time increase than for the case of $p \leq 23$, but it is lesser than the conventional endomorphism based architecture.

Computational Time Analysis

In [13] the ALU computes the latencies as follows : the multiplication requires 63 clock cycles, the addition and subtraction module requires 18-19 clock cycles, the division module requires 17 clock cycles if the least significant bit is zero or its one. So, the division's average latency is 16 clock cycles. The inversion module takes 13640 clock cycles. Computing the point doubling requires 280 clocks cycles in our architecture, the multiplication requires 151.2 ns computational time, the addition and subtraction module requires a total of 83 ns. The LUT main requires 1.26 ns, the mux requires total 1. 5ns.The overall latency of the architecture is 525.626ns.

The computational analysis of the architecture is as follows

Table 5. Computational time analysis of modules

Module	[12]	Proposed Architecture
Endomorphism	$k = k_1 + k_2\lambda$	$k = k_1 + k_2\lambda_1 + k_3\lambda_2$
Multiplication	198ns	151.2ns
Adder	78ns	83ns
LUT	955ns	1.26ns
Slices	377	85
LUT-Gen	Offline	Online
Total compilation	1345ns	525.64ns

MEMORY UTILIZATION

The memory utilization is the main concern when it comes to improve the efficiency of a system. For making a system more efficient the memory utilization is optimized. In [13], a lookup table with extended offline points has to be stored in the RAM. The double scalar multiplication requires an execution time of 415392 clock cycles and the RAM consumption is 12222 bytes and thus sacrificing 12% performance.

As a comparison, the double scalar multiplication without using endomorphism i.e., by only using interleaving method requires an execution time of 454,179 clock cycles. This result shows that endomorphism speeds up the system by 8.5-19%.

Table 6. Memory slices comparison with [12]

Memory	Proposed architecture	[12]
Total slices utilized	85 slices	377 slices

Comparison with Other Implementations

The results show a comparison of our proposed architecture with other related works. There are no implementations in literature with the same security level as ours, hence a direct comparison is not possible, but we have highlighted the size of prime fields.

In this research, double scalar multiplication is utilized for ECDSA signature verification, with a scalable base point and public key. Only 4 points are stored from pre-computation, as an efficient endomorphism is being used to accelerate the computation of $k.G$. The ECDSA signature generation requires a constant time inversion that has been done using the Fermat's theorem in [13] by using a 257 modular squaring and 14 modular multiplications. As most of the previous techniques only report the signature generation time of ECDSA algorithms, the verification time is estimated by simply multiplying the generation time by two.

The authors of [22] and [23] have optimized their architecture using a microcode structure, but their implementation requires extra instruction decoding modules and has high ROM consumption. This research does not require extra hardware for memory as it can generally lookup tables (LUTs) on run time.

Discussion

As described earlier the double scalar multiplication in ECDSA verification is a time-consuming process, therefore, it requires an efficient implementation and optimization techniques. The double scalar multiplication is of the form producing the sum of two scalar products. G is fixed and Q is an arbitrary point on the curve [13], some fixed value of G is taken to optimize the overall system performance. In this research the core advantage lies in the fact that G is not fixed, providing a pre-eminent advantage that the user's public key can change at run time. The lookup tables are generated online, thus the latency is increased for the only first time then the lookup tables will be used to calculate points till $p \leq 257$. The throughput of the system increases as a result.

The proposed architecture does not require extra architectural blocks in comparison with the latest research [13], which requires 20 DSP 48E1 blocks that become the bottleneck for the system.

CONCLUSION AND FUTURE WORK

Elliptic Curve Cryptography provides the security needed for wireless communication, the major drawback of ECC is the verification time of the messages containing the elliptic curve digital signature algorithm. The double scalar multiplication being the root cause affects the overall performance of the system at the receiver side.

The prime focus of work presented in this chapter is on improving the signature verification efficiency of the receiver side, by improving the time taken by the double scalar multiplication. By introducing a multilevel endomorphism, not only the double scalar multiplication time has been improved but the hardware requirements of the system have also been reduced. The architecture is designed for $p \leq 23$, which can be extended for any value of prime field Fp.

As in [13], the solution for the double scalar problem is suggested, but the base point G is kept constant. Thus, the system is restricted to be specific for a certain public key. The proposed architecture is generic and works for any public key. Look-up tables are generated online, thus improving the throughput of the overall system.

For future work, more levels of endomorphism can be introduced, thus further reducing the computational time. Multiple architectures can be proposed for the double scalar multiplication time reduction by further investigating the pre-computation and scalar division (endomorphism). However, multiple other techniques of parallel processing and pipelining can be utilized to improve the verification time, thus making the system faster and more efficient.

REFERENCES

[1] L. Atzori, A. Iera, and G. Morabito, "The Internet of Things: A survey," *Comput. Networks*, vol. 54, no. 15, pp. 2787–2805, 2010.

[2] M. A. Javed, E. Ben Hamida, and W. Znaidi, "Security in intelligent transport systems for smart cities: From theory to practice," *Sensors (Switzerland)*, vol. 16, no. 6, 2016.

[3] P. Demo, "Secure Communication in Vehicular Networks," ... *IEEE Veh. Netw. Conf. Demo ...*, 2012.

[4] C. Jeong and Y. Kim, "Implementation of efficient SHA-256 hash algorithm for secure vehicle communication using FPGA," *ISOCC 2014 - Int. SoC Des. Conf.*, pp. 224–225, 2015.

[5] V. Nikov and P. Rombouts, "*Low-Latency ECDSA Signature Verification* —," vol. 24, no. 11, pp. 3257–3267, 2016.

[6] Y. E. Krasteva, J. Portilla, E. De La Torre, and T. Riesgo, "Embedded runtime reconfigurable nodes for wireless sensor networks applications," *IEEE Sens. J.*, vol. 11, no. 9, pp. 1800–1810, 2011.

[7] C. Paper, "*System Modeling and Fault Analysis of a HVAC Transmission Line*," no. May 2015, 2014.

[8] J. Silverman, "Fast multiplication in finite fields GF (2N)," *Ches*, pp. 122–134, 1999.

[9] R. Afreen and S. C. Mehrotra, "A Review of Elliptic Curve Cryptography for Embedded Systems," *Int. J. Comput. Sci. Informatiion Technol.*, vol. 3, no. 3, pp. 84–103, 2011.

[10] W. Stallings, *Cryptography and Network Security: Principles and Practices 4th edition*. 2005.

[11] J. Andersen and S. Sutcliffe, "Intelligent Transport Systems (ITS) - An Overview," *IFAC Proc. Vol.*, vol. 33, no. 18, pp. 99–106, 2000.

[12] H. Leung, N. E. El Faouzi, and A. Kurian, "Intelligent transportation system (ITS)," *Inf. Fusion*, vol. 12, no. 1, pp. 2–3, 2011.

[13] Z. Hu, J. Kimmo, Z. Liu, J. Großsch, H. Wang, and I. Verbauwhede, "*Elliptic Curve Cryptography with Efficiently*

Computable Endomorphisms and Its Hardware Implementations for the Internet of Things.pdf," vol. 66, no. 5, pp. 773–785, 2017.

[14] J. A.solnas, "*Low-weight binary representations for pairs of integers,*" 2013.

[15] M. F. Siddiqui, A. W. Reza, J. Kanesan, and H. Ramiah, "Investigation of a novel common subexpression elimination method for low power and area efficient DCT architecture," *Sci. World J.*, vol. 2014, 2014.

[16] I. Engineering, "*ECC encryption system using encoded multiplier and Verdic mathematics*," pp. 5531–5538, 2013.

[17] I. High-speed, D. S. Processing, F. P. Multipliers, and W. S. Network, "*Chapter 3 Design of optimized elliptic curve*," pp. 28–58.

[18] J. Xie, P. K. Meher, and Z. H. Mao, "High-throughput finite field multipliers using redundant basis for FPGA and ASIC implementations," *IEEE Trans. Circuits Syst. I Regul. Pap.*, vol. 62, no. 1, pp. 110–119, 2015.

[19] S. Gao, P. Langlois, D. Al-Khalili, and N. Chabini, "Asymmetric large size multipliers with optimised FPGA resource utilisation," *IET Comput. Digit. Tech.*, vol. 6, no. 6, pp. 372–383, 2012.

[20] I. Setiadi, A. I. Kistijantoro, and A. Miyaji, "Elliptic curve cryptography: Algorithms and implementation analysis over coordinate systems," *ICAICTA 2015 - 2015 Int. Conf. Adv. Informatics Concepts, Theory Appl.*, no. November, 2015.

[21] D. Hankerson, A. Menezes, and S. Vanstone, "*Guide to Elliptic Curve Cryptography*," 2004.

[22] M. Hutter, M. Feldhofer, and T. Plos, *An ECDSA processor for RFID authentication*. 2010.

[23] E. Wenger, M. Feldhofer, and N. Felber, "Low-resource hardware design of an elliptic curve processor for contactless devices," *Lect. Notes Comput. Sci. (including Subser. Lect. Notes Artif. Intell. Lect. Notes Bioinformatics)*, vol. 6513 LNCS, pp. 92–106, 2011.

In: Vehicular Ad-Hoc Networks
Editors: G. G. Md. Nawaz Ali et al.
ISBN: 978-1-53618-038-1

Chapter 3

EXPLOITATION OF NETWORK CODING IN ON-DEMAND WIRELESS DATA BROADCAST

G. G. Md. Nawaz Ali*[1,*]*, Victor C. S. Lee*[2]*, Yuxuan Meng*[2]*,
***Peter H. J. Chong*[3] *and Jun Chen*[4]**
[1]Department of Applied Computer Science,
University of Charleston, WV, US
[2]Department of Computer Science, City University of Hong Kong,
Kowloon, Hong Kong
[3]Department of EEE, Auckland University of Technology,
Auckland, New Zealand
[4]School of Information Management, Wuhan University,
Wuhan, China

Abstract

On-demand broadcast is a scalable approach to disseminating information to a large population of clients while satisfying dynamic needs of clients, for instance in vehicular networks. However, in conventional broadcast approaches, only one data item can be retrieved by clients in one broadcast tick. To further improve the efficiency of wireless bandwidth, in this work, we give a comprehensive study on incorporating network coding with representative on-demand scheduling algorithms while preserving their original scheduling criteria. In particular, a graph model

*Corresponding Author's Email: ggmdnawazali@ucwv.edu.

is derived to maximize the coding benefit based on the clients' requested and cached data items. Further, we propose a heuristic coding based approach, which is applicable for all the on-demand scheduling algorithms with low computational complexity. We give a running example demonstrating how the proposed algorithm works for representative algorithms. In addition, based on various application requirements, we classify the existing on-demand scheduling algorithms into three groups, real-time, non-real-time and stretch optimal. In view of different application-specific objectives, we implement the coding versions of representative algorithms in each group. Extensive simulation results conclusively demonstrate the superiority of coding versions of algorithms against their non-coding versions on achieving their respective scheduling objectives.

Keywords: on-demand broadcasting, network coding, scheduling, data broadcasting

1. INTRODUCTION

Data broadcast is an attractive solution for large scale data dissemination in wireless communication environments. With ever expanding of data-centric mobile applications and the recent advances in wireless communication technologies, data broadcast techniques have attracted much attention in a variety of mobile computing environments [1, 2, 3]. In general, there are two types broadcast approaches: push-based broadcast and pull-based broadcast [4, 5]. In push-based broadcast, the scheduler periodically disseminates data items based on the historical data access patterns of clients. This approach is suitable for applications with small database sizes and stable data access patterns, such as emergency message broadcast from the backend server [6, 7]. On the other hand, in pull-based broadcast, which is commonly known as on-demand broadcast, the scheduler disseminates data items based on explicit requests received from clients, such as traffic update query in traffic navigation systems [8, 9, 10]. In other words, the scheduling decisions are made on-line in on-demand broadcast, which is more suitable for dynamic and large-scale data dissemination services [11]. In this work[1], we focus on further improving the bandwidth efficiency of on-demand broadcast by exploiting network coding.

[1]This work is an extend version of our work [12]. Some figures are reproduced with the proper permission from MDPI.

A great number of on-demand scheduling algorithms have been proposed in previous studies [11, 13, 14], which have different application-specific performance objectives. Consider a broadcast tick as the duration for transmitting one unit-sized data item. All these strategies can only broadcast one data item in a broadcast tick. This restriction hinders the further improvement of broadcast bandwidth efficiency. In contrast, network coding [15] enables the broadcast of multiple data items via an encoded form in a broadcast tick. We observe that an inherited advantage of on-demand broadcast is having the feedback information from clients to the server, and hence applying network coding in on-demand broadcast does not require additional overhead for explicit feedback.

A variety of network coding techniques have been studied. Random linear network coding (RLNC) [16] is suitable for push based broadcast, specially for broadcasting popular contents which may serve a large population of clients [6, 17], such as periodic safety message, traffic information, etc. However, RLNC is not adaptive to customized services, such as mobile infotainment, gas station query, parking space query, etc. To support such services, the server needs to receive explicit on-demand requests from clients, so that it can provide services by encoding corresponding data items accordingly. XOR ($\oplus$)-based network coding is a suitable candidate for such cases, as it has very low encoding and decoding overhead and it is simple to be implemented [18]. Many studies have applied XOR-based coding for improving network throughput [18, 19, 20] and enhancing packet delivery ratio [21, 22].

With respect to the different performance objectives, the on-demand scheduling algorithms can be classified into three groups: *real-time*, *non-real-time* and *stretch optimal* scheduling algorithms. Specifically, in real-time scheduling, the primary objective is to minimize the deadline miss ratio, namely, serving as many requests as possible before their deadlines expire. For instance, in vehicular networks, real-time traffic information has to be broadcast to vehicles with certain time-constraint (e.g., must serve before a vehicles exits the service range of a road-side unit) [23, 24]. Representative real-time scheduling algorithms include EDF (Earliest Deadline First) [25] and SIN (Slack time Inverse Number of pending requests) [11]. In non-real-time scheduling, the primary objective is to minimize the average response time, namely, serving the requests as soon as possible. For instance, mobile clients wish to receive non-real-time infotainment information as soon as possible [13]. Representative non-real-time scheduling algorithms include FCFS (Fist Come First Served) [26], MRF (Most Requested First) [27], LWF (Longest Wait First) [27] and R $\times$W (Number of

Pending Request Multiply Waiting Time) [4]. In stretch optimal scheduling, the primary objective is to minimize the stretch, which is the ratio of the response time to the service time, where the service time refers to the transmission time of the data items. For instance, fairness treatment to different size requests in satellite networks, wireless LANs, celluar networks etc. [28]. Representative stretch optimal scheduling algorithms include LTSF (Longest Total Stretch First) [29] and STOBS (Summary Table On-demand Broadcast Scheduler) [28].

Although existing studies have considered to improve scheduling performance with respect to different application-specific objectives, they do not fully utilize the advantage of on-demand broadcast, where the server is able to be aware of the requested and cached data items of mobile clients. The impact of network coding on different scheduling objectives highly depends on the applied coding strategy at the server. In order to evaluate possible benefits brought by network coding, apparently, it is desirable to preserve the original scheduling criteria when applying network coding to existing on-demand scheduling algorithms. In this work, we will investigate the potential benefit of applying network coding into different groups of on-demand scheduling algorithms. The major contributions of this work are stated as follows.

- We present a graph model, where a CR-graph (Cache-Request-graph) [14, 30] is formed based on the clients requested and cached data items for reducing redundant broadcasting.
- We transform the scheduling problem to the problem of finding maximum clique in the derived CR-graph, which is a well-known NP-complete problem in graph theory [31, 32]. Accordingly, we propose a heuristic coding based approach, which ensures that the each scheduling algorithm will transform into its corresponding coding version while preserving its original scheduling criterion.
- For the three groups of scheduling algorithms, namely, real-time, non-real-time and stretch optimal, we select representative solutions in each group and present the detailed implementation of their coding versions with the proposed approach, followed by a running example.
- In order to evaluate the impact of network coding to different groups of algorithms, we build the simulation model for each group and design the performance metrics.

- We give a comprehensive performance evaluation, which demonstrates the efficiency of network coding on enhancing system performance with respect to different scheduling objectives under various circumstances.

The rest of the chapter is organized as follows. Section 2 reviews the related work. Section 3 presents the system model. Section 4 proposes a heuristic coding based approach followed by a running example. We build the simulation model in Section 5. An extensive performance evaluation is conducted in Section 6. Finally, we conclude this work in Section 6.3.4.

2. Related Works

The two main components of the proposed work in this chapter are the on-demand wireless data broadcast and network coding. In the following subsections, first we review some existing representative on-demand scheduling algorithms in real-time, non-real-time and stretch optimal models. Then we review some exiting works in network coding, and state the possibilities and challenges of applying network coding in on-demand wireless data broadcast.

2.1. Real-Time Model

In real-time applications, the requests submitted by clients are associated with deadlines, and the primary objectives of real-time scheduling is to minimize the deadline miss ratio of requests. EDF (Earliest Deadline First) [25] is one of the foremost classical scheduling algorithms in real-time systems. It broadcasts the data item with the shortest remaining lifetime to cater for the urgency of requests. Hu [33] considered request urgency, service productivity and scheduling fairness in algorithm designing. The objective is to achieve low request deadline miss rate and mean access time simultaneously. Xu et al. [11] proposed scheduling algorithm for time-critical requests called SIN (Slack time Inverse Number of pending requests). It is motivated by two existing strategies: EDF, which considers the urgency of requests, and MRF, which involves the productivity of data broadcast. SIN outperforms other existing real-time on-demand broadcast algorithms significantly. Chung et al. [34] gave a theoretical analysis on the minimum required number of channels to complete time-critical services given a certain time-bound. An algorithm called SUSC (Scheduling under Sufficient Channels) is proposed to generate the broadcast program when there are sufficient channels. Alternatively, an algorithm called PAMAD (the Progressively Approaching Minimum Average Delay) is proposed to handle the situation where the number of channels is insufficient. Chen et al. [35] focused on scheduling for multi-item requests with time constraints in on-demand broadcast environments. A algorithm called DTIU (Dynamic Temperature Inverse Urgency) is proposed, which makes scheduling decisions by considering both request popularity and request timing requirements. Ali et al. [36] studied the performance of different scheduling algorithm under strict deadline environment. He et al. [37] proposed UPF (Urgent and Popular data item First) for scheduling real-time requests under multi-channel environment.

2.2. Non-Real-Time Model

In non-real-time applications, the responsiveness of the system is measured by the time from the submission of a request to the time that this request has been served. The primary objectives of non-real-time scheduling algorithms is to minimize the response time of serving requests. Many classical algorithms have been proposed for non-real-time applications. FCFS (First Come First Served) [26] broadcasts data items sequentially according to their arrival order. SRPT (Shortest Remaining Processing Time) [29] chooses the data item with the shortest service time to broadcast. Wong [27] proposed two well-known strategies for scheduling requests in on-demand broadcast systems: MRF (Most Requested First) and LWF (Longest Wait First). MRF broadcasts the data item that has the largest number of pending requests to account for broadcast productivity. LWF calculates the total time that all pending requests for a data item have been waiting. The data item with the longest total waiting time is chosen to broadcast. Aksoy and Franklin [4] pointed out that the straightforward implementation of LWF is expensive and not practical for large systems. They proposed a low-overhead and scalable scheduling algorithm called R×W (Number of Pending Requests Multiply Waiting Time) that combines the benefits of MRF and FCFS. It calculates the number of pending requests for a data item multiplied by the amount of time that the oldest outstanding request for that data item has been waiting. At each broadcast tick, the request with the maximum R×W value will be chosen to serve. Liu and Lee [5, 13] analyzed the mean response time of different existing scheduling algorithms under multi-channel environment. Polatoglou et al. [38] studied the complexity of adaptive push based broadcast system for reducing mean response time. Lu et al. [39] studied the combined data push and data retrieval scheduling problem in multi-channel environment.

2.3. Stretch Optimal Model

In stretch optimal scheduling, the algorithm takes the variable sizes of data items into consideration (i.e., the service time to different data items). A metric called *stretch* is proposed in [29]. It is the ratio of the response time to the service time of the requested data item. The response time refers to the duration from the instance when the request is submitted to the time when the request is satisfied, whereas the service time is the duration for transmitting the data item. An algorithm called LTSF (Longest Total Stretch First) is proposed in [29] to broadcast the data item with the largest current stretch, where the current stretch of a data item is the ratio of the aggregated waiting time of all the outstanding requests for the data item to the service time of the data item. Wu and Cao [40] reduced the computation overhead to make the stretch-based scheduling more scalable in on-demand broadcast environments. Sharaf and Chrysanthis [28] proposed an algorithm called STOBS (Summary Tables On-demand Broadcast Scheduler), which broadcasts the data item with the maximum $\frac{R \times W}{S}$ value, where R, W and S represent the number of pending requests for the data item, the waiting time of the oldest request and the item size, respectively. Lee et al. [41] proposed PRDS (Preemptive Request Deadline Size), for minimizing the response time and stretch in scheduling size variant data item.

2.4. Network Coding

Random linear network coding (RLNC) is initially proposed for the transmission and compression of information in multi-source multi-cast networks [16]. Ye et al. [6] applied RLNC for pushing popular contents from a single sever to multiple clients. Hassanabadi and Valaee [17] proposed to apply RLNC for repetitive reliable safety message rebroadcasting in the congested wireless channel. Ploumidis et al. [42] analyzed the throughput and delay in wireless mesh networks incorporating RLNC. Qu et al. [43] also studied the trade-off between the throughput and decoding delay in applying RLNC in wireless mesh networks, and proposed a protocol named DCNC (Delay Controlled Network Coding). However, as in RLNC, the random coefficients need to be attached with the coded message, the coding overhead can be significant in a congested network. To reduce the overhead, a number of studies have considered XOR-based network coding in on-demand broadcast. Chu et al. [44] proposed On-demand

Encoding (OE) algorithm for reducing clients' access time in on-demand broadcast. However, OE can only encode two data items in an encoded packet, and hence it is not flexible and it cannot maximize the coding benefit. Chaudhry and Sprintson [30] transformed the on-demand broadcast coding problem to the boolean satisfiability (SAT) problem. They proposed several heuristic solutions based on graph coloring and color saving algorithms. Gao et al. [45] studied XOR coding for packet retransmission in erroneous wireless medium. Ma et al. [46] did the initial investigation of impact of network synchronization on the integration of scheduling and network coding. They proposed a strategy called COSE (Combined Opportunistic Scheduling and Encoding). Chen et al. [47] proposed an on-demand broadcast algorithm called ADC (Adaptive Demand-oriented Coding) for serving multi-item requests. In [14], Chen et al. proposed ADC-1 and ADC-2, where ADC-1 considers scheduling and network coding operations separately, and in ADC-2, these two operations are considered integrally. Zhan et al. [20] proposed a model for constructing encoding packet with a number of data items. The proposed approach greedily finds maximal cliques based on constructed graph model to find an approximate solution to the formulated coding problem.

In summary, to improve the scheduling performance by further exploring the strength of on-demand broadcast, it is desirable to adopt network coding into data broadcast. With network coding, different data items can be encoded into one packet. Since different coding strategies have different impacts on scheduling performance, in this chapter, we propose a general heuristic coding based on-demand broadcast scheme, by which the data scheduling problem can be transformed to the maximum clique problem. With the proposed approach, existing on-demand scheduling algorithms can be migrated into their corresponding coding versions while preserving their original criteria in scheduling data items. In addition, this work validates the advantage of applying network coding in on-demand scheduling through a comprehensive simulation study.

3. System Model

3.1. System Architecture

The system architecture shown in Figure 1 represents a typical on-demand data broadcast system in wireless communication environments [4]. The system consists of one server and a number of clients. When a data item is requested by a client but it is not in the local cache, the client submits a request via the uplink channel to the server, and monitors the downlink channel waiting for the requested data item. Typically, the bandwidth of the downlink channel is much greater than that of the uplink channel, which exhibits an asymmetric wireless communication environment. According to a certain scheduling algorithm, the server retrieves the requested data items from the local database and constructs an encoded packet for broadcasting. The client gets the requested data item by decoding the received packet, and then the decoded data item will be cached locally. A client can decode a requested data item from an encoded packet if it caches all the other data items in the encoded packet except the requested one. The XOR operation is commonly adopted for encoding and decoding due to its trivial computation overhead [14, 44, 48]. For example, as shown in Figure 1, C3 (i.e., the Client 3) has cached d_1. Suppose C3 requests d_2 and the server broadcasts an encoded packet $d_1 \oplus d_2$. C3 is able to decode $d_1 \oplus d_2$ and retrieve d_2 by such a schedule. Due to the limited cache size of clients, the LRU (Least Recently Used) cache replacement policy is adopted for cache management. A client will generate a new request only when the previous submitted request is satisfied or misses its deadline, which resembles the closed system model [14, 49].

3.2. Graph Model

The key to improve the system performance for real-time, non-real-time and stretch optimal models is to disseminate more number of requested data items per broadcast tick. From the network coding perspective, needs to make the encoding decision in such a way so that more number of clients will be satisfied per coded packet and eliminates the redundant encoding. For this, assume that $G(V,E)$ is a CR-graph (Cache-Request-Graph) which is formed from the clients' cached and requested data items. The next step is to find a maximum clique which contains the maximum priority vertex based on the used scheduling algorithm. In on-demand broadcast, CR-graph construction is a practical assumption because clients can piggyback their updated cache information in the requests submitted to the server. With the knowledge of clients' cached and requested items, a graph $G(V,E)$ is constructed at the server to guide the encoding decision. A brief overview of graph construction is given below. The summary of used notations in this chapter is shown in Table 1.

The set of clients in the system is represented by $C = \{c_1, c_2, \cdots, c_n\}$,

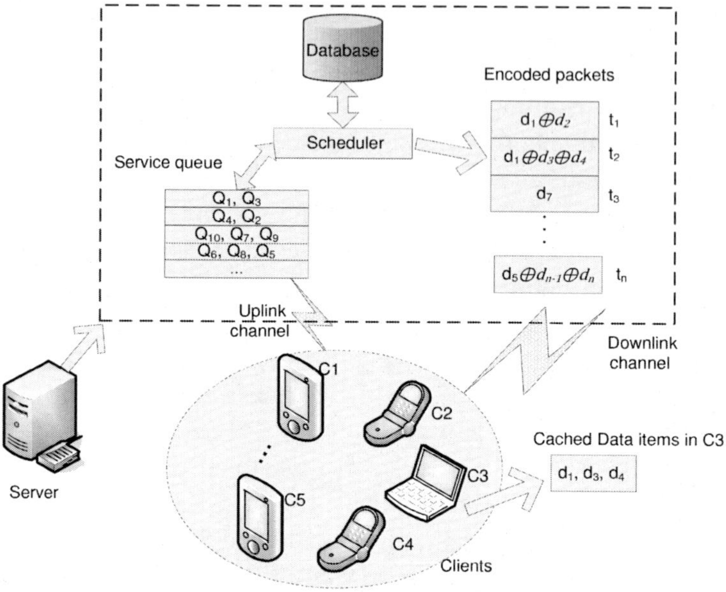

Figure 1. System architecture.

where n is total number of clients. $R(c_i)$ represents the set of data items requested by client c_i, and $CH(c_i)$ represents the set of data items cached at client c_i ($1 \leq i \leq n$). The server maintains a database D with m data items, where d_j is the jth data item ($1 \leq j \leq m$).

Definition 1. *Based on the clients' requested and cached data item, a graph* $G(V,E)$ *is constructed by the following rules:*
$V = \{v_{ij} |\ client\ c_i\ requests\ for\ item\ d_j,\ 1 \leq i \leq n,\ 1 \leq j \leq m\}$
$E = \{(v_{i_1 j_1}, v_{i_2 j_2}) |\ j_1 = j_2;\ or\ j_1 \neq j_2,\ d_{j_2} \in CH(c_{i_1}),\ d_{j_1} \in CH(c_{i_2})\}$

Accordingly, the rules for constructing edges are specified below:

- $e(v_{i_1 j_1}, v_{i_2 j_2})$ with $j_1 = j_2$ means that if client c_{i_1} and client c_{i_2} request the same data item, there is an edge between the two vertices $v_{i_1 j_1}$ and $v_{i_2 j_2}$.
- $e(v_{i_1 j_1}, v_{i_2 j_2})$ with $j_1 \neq j_2$, $d_{j_2} \in CH(c_{i_1})$, and $d_{j_1} \in CH(c_{i_2})$ means that if client c_{i_1}'s cache contains the data item being requested by client c_{i_2} and vice versa, there is an edge between vertices $v_{i_1 j_1}$ and $v_{i_2 j_2}$.

A clique δ is a subset of the vertices in the CR-graph G, such that every two vertices in the subset are connected by an edge, i.e., $\delta \subseteq V(G)$.

Table 1. Summary of notations

Notation	Description	Notes
c_i	A client	$c_i \in C; C = \{c_1, c_2, \cdots, c_n\}$
d_j	A data item	$d_j \in D; D = \{d_1, d_2, \cdots, d_m\}$
$R(c_i)$	Set of requested data items by c_i	
$CH(c_i)$	Set of cached data items by c_i	$R(c_i) \cap CH(c_i) = \emptyset$
$G(V,E)$	A graph	
v_{ij}	A vertex representing client c_i requests data item d_j	$v_{ij} \in V(G)$, $1 \le i \le n$, $1 \le j \le m$
$N(d_j)$	popularity of d_j	
$d_{j(i)}$	Client c_i is pending for d_j	
t_i	Waiting time of c_i	
$\sum t(d_j)$	Summed waiting time of clients pending for d_j	
T_i	Slack time of c_i	
l_i	Size of the requested data item of c_i	
δ	A clique	
$\|\delta\|$	Number of vertices in δ	
C_δ	Set of clients covered in δ	$C_\delta = \{c_1, c_2, \cdots, c_{\|\delta\|}\}$
$\beta(\delta)$	Encoded packet for the clique δ	$\beta(\delta) = \{d_1 \oplus d_2 \oplus \cdots \oplus d_K\}$
K	Number of different data items encoded in $\beta(\delta)$	$K \le \|\delta\|$
δ_{max}	Maximum clique	
$\delta_{max}^{v_{ps}}$	Maximal clique covering vertex v_{ps}	$\|\delta_{max}^{v_{ps}}\| \le \|\delta_{max}\|$
B	Channel broadcast bandwidth	

Definition 2. *Let an arbitrary clique δ contains $|\delta|$ number of vertices, i.e., $\delta = \{v_{i_1 j_1}, v_{i_2 j_2}, \cdots, v_{i_{|\delta|} j_{|\delta|}}\}$. The corresponding clients are $C_\delta = \{c_1, c_2, \cdots, c_{|\delta|}\}$. The encoded packet for clique δ will be $\beta(\delta) = \{d_1 \oplus d_2 \oplus \cdots \oplus d_K\}$, where K different data items are requested by clients in C_δ. Note that $K \le |\delta|$.*

By encoding the set of data items in δ for broadcasting, scheduling with coding can satisfy multiple requests for different data items in a broadcast tick. In contrast, scheduling without coding can only serve the requests for the same data item in a broadcast tick.

3.3. Coding Based On-Demand Broadcast

Recalling v_{ij} denotes a client c_i for the requested data item item d_j. For simplicity, we assume that the vertices in the graph are denoted by $v_1, v_2, \cdots, v_n$, where n is number of vertices. Note that this simple assumption does not refrain a vertex to also denote the corresponding requested data item. v_i corresponds to the request c_i, where $v_i \in V(G)$ and $1 \leq i \leq n$. A vertex v_i is associated with three weights, namely t_i, l_i, *and* T_i, which represent the waiting time, the item size and the slack time, respectively. A data item d_j is also associated with a weight $N(d_j)$, which represents the popularity, namely the number of pending requests for d_j. The broadcast channel bandwidth is B. The relationships between the scheduling objectives and the properties of the cliques are stated as follows.

- Relationship 1: Real-time scheduling algorithms which aim to serve the request with the minimum slack time such as EDF, should select the largest clique δ which includes the vertex v_i with the largest $\frac{1}{T_i}$ value. The other real-time algorithm SIN, also considers popularity in addition of slack time.

- Relationship 2: Non-real-time scheduling algorithms which aim to serve the request with the longest waiting time such as FCFS, should select the largest clique δ which includes the vertex v_i with the largest t_i value.

- Relationship 3: Non-real-time scheduling algorithms which aim to broadcast the data item with the highest popularity such as MRF, should select the largest clique δ which includes the maximum number of vertices.

  The working properties of other non-real-time algorithms are the combination of the properties of FCFS (waiting time) and MRF (data item popularity), such as LWF and R×W, both consider waiting time and data item popularity for scheduling.

- Relationship 4: Stretch optimal scheduling algorithms which aim to reduce the stretch such as LTSF, should select the clique δ which has the maximum current stretch, namely with the maximum summation of $\frac{t_i}{\frac{l_i}{B}}$ value. The other stretch optimal scheduling algorithm STOBS considers data item popularity in addition of request waiting time and data item size.

The relationships of all the algorithms are summarized in Table 2. From the above relationships it is understandable that to migrate the traditional scheduling algorithms into the network coding versions by exploiting the full coding benefits, we need to find the maximum clique δ_{max} in the graph $G(V,E)$. This is because the number of satisfying clients is equivalent to the number of vertices

Table 2. Coding based on-demand broadcast

Type	Algorithm	Remarks
Real-time	EDF	Find the largest clique δ with T_{min}
	SIN	Find the largest clique δ with the maximum $\frac{\|\delta\|}{T_{min}}$ value
Non-real-time	FCFS	Find the largest clique δ with t_{max}
	MRF	Find the largest clique δ
	LWF	Find the largest clique δ with the maximum summed waiting time
	$R \times W$	Find the largest clique δ with the maximum $\|\delta\| \times t_{max}$ value
Stretch optimal	LTSF	Find the largest clique δ with the maximum $\frac{\sum_{v_i \in \delta} t_i}{max_{v_i \in \delta}\{\frac{l_i}{B}\}}$ value
	STOBS	Find the largest clique δ with the maximum $\frac{\|\delta\| * max_{v_i \in \delta} t_i}{max_{v_i \in \delta} l_i}$ value

in δ_{max}. However finding the maximum clique is a well-known NP-complete problem in graph theory [31]. This is computationally expensive, because the new maximum clique needs to find out every time before making a broadcast decision. This turns out applying network coding in on-demand broadcast may be infeasible in practice. Hence in the following, we propose a heuristic coding based approach to convert the traditional scheduling algorithms into the coding version in polynomial time, while preserving the original scheduling criteria.

4. Proposed Heuristic Coding Based On-Demand Broadcast

Recalling the objective of this work is to convert a traditional scheduling algorithm into the coding version with its original scheduling criteria along with maximizing the coding benefit by keeping the overhead minimal. Hence the key idea of the proposed heuristic coding based approach is to broadcast the encoded packet, which (1) must contain the vertex corresponds to the scheduled client by the scheduling algorithm, and (2) satisfy as many clients as possible per broadcast by keeping the encoding complexity as minimal as possible. The proposed approach contains the following two important considerations.

- Find the vertex, which holds the maximum priority value of the scheduling algorithm. If a data item $d_s \in D$ of a client c_p is selected to broadcast, the corresponding vertex is v_{ps}, we call as *selected vertex*.
- To maximizing the exploitation of network coding and reducing encoding complexity, need to find the maximal clique $\delta_{max}^{v_{ps}}$ (approximate maximum

clique) which covers the selected vertex v_{ps}. Note that $\delta_{max}^{v_{ps}}$ is the maximum among all the possible maximal cliques covering v_{ps} in G. $\delta_{max}^{v_{ps}}$ will be used to form the encoded packet $\beta(\delta_{max}^{v_{ps}})$ for broadcasting.

Lemma 1. *For an arbitrary vertex $v_{ij} \in \delta_{max}^{v_{ps}}$, the corresponding client $c_i \in C_\delta$ can retrieve d_j from the encode packet $\beta(\delta_{max}^{v_{ps}})$.*

Proof. As the proposed system is a closed system, which infers that one client can have only one pending request in the service queue. Hence, we can infer that all the vertices in a clique represents the unique clients. According to the clique definition, all the vertices in a clique are inter-connected. Again according to the edge definition of a CR-graph (Definition 1), a client only has edges with other clients if they request the same data item, or one client's cached data item is being requested by another client and vice versa. This implies that for a client $c_i \in C_\delta$, if $v_{ij} \in \delta_{max}^{v_{ps}}$, for $K \geq 1$, c_i already has (in the cache) all the data items of $\beta(\delta_{max}^{v_{ps}})$, except d_j; otherwise all the vertices in $\delta_{max}^{v_{ps}}$ requested for the same data item d_j and $d_j = d_s$. Hence c_i can easily decode d_j from $\beta(\delta_{max}^{v_{ps}})$ with its cached data items. □

Lemma 2. *Broadcasting the encoded packet $\beta(\delta_{max}^{v_{ps}})$ for the selected vertex v_{ps}, ensures the migration of a scheduling algorithm into the coding version with preserving its original scheduling criteria.*

Proof. The proposed coding based system first select the selected vertex v_{ps}, then finds the maximal clique $\delta_{max}^{v_{ps}}$, which covers v_{ps}. Note that v_{ps} is the highest priority vertex based on the scheduling criteria of a scheduling algorithm. For instance, in the real-time model, for EDF scheduling, v_{ps} is vertex with the highest urgency; in the non-real-time model, for FCFS scheduling, v_{ps} is vertex with the longest waiting time; in the stretch optimal model, for LTSF scheduling, v_{ps} is vertex with the longest stretch. After selecting v_{ps}, the maximal clique $\delta_{max}^{v_{ps}}$ is found from the CR-graph G, which cover v_{ps}. This means that encoded packet $\beta(\delta_{max}^{v_{ps}})$ must contain the selected data item d_s of client c_p. Hence by broadcasting $\beta(\delta_{max}^{v_{ps}})$, all the scheduling algorithms' criteria are preserved in the coding version. □

4.1. The Heuristic Algorithm

To sum up, our heuristic coding based on-demand broadcast consists of the following three key steps. The pseudo code is shown in Algorithm 1.

Algorithm 1: Heuristic coding based on-demand broadcast.

Step 1. CR-graph construction
$G(V,E)$ is the already constructed CR-graph, and v_{kr} is the vertex for a newly arrived client c_k, where, $1 \leq k \leq n$, $1 \leq r \leq m$;
for *each* $v_{ij} \in V(G)$ **do**
    if j=s **then**
        $E(G) \leftarrow E(G) + e(v_{ij}, v_{kr})$;
    else if $d_j \in H(c_k)$ & $d_r \in H(c_i)$ **then**
        $E(G) \leftarrow E(G) + e(v_{ij}, v_{kr})$;
    else
        /* There is no link between v_{ij} and v_{kr} */
$V(G) \leftarrow V(G) + v_{kr}$;
Step 2. Find the selected vertex v_{ps}
Each vertex $v_{ij} \in V(G)$ is associated with T_i, t_i, l_i;
Each data item $d_j \in D$ is associated with $N(d_j)$;
Initialize $MAX \leftarrow 0$;
for *each* $v_{ij} \in V(G)$ **do**
    $P \leftarrow$ Invoke the underlying scheduling algorithm with T_i, t_i, l_i, $N(d_j)$;
    if $P > MAX$ **then**
        $MAX \leftarrow P$;
        $v_{ps} \leftarrow v_{ij}$;
Step 3. Encode and broadcast the encoded packet $\beta(\delta_{max}^{v_{ps}})$
Find the maximal clique $\delta_{max}^{v_{ps}}$ for the selected vertex v_{ps};
Form the encoded packet $\beta(\delta_{max}^{v_{ps}}) \leftarrow \{d_1 \oplus d_2 \oplus \cdots \oplus d_k\}$; $k \leq |\delta_{max}^{v_{ps}}|$;
Broadcast $\beta(\delta_{max}^{v_{ps}})$;
/* Update $G(V,E)$ */
for *each* $v_{ij} \in \delta_{max}^{v_{ps}}$ **do**
    $V(G) \leftarrow V(G) - v_{ij}$;

Step 1. CR-graph $G(V,E)$ construction

Upon a request arrival from a client c_k for the data item d_r, the corresponding vertex v_{kr} will be added in the vertex set $V(G)$. According the edge definition as defined in Definition 1, all the possible edges are formed for v_{kr} with the vertices in $V(G)$.

Step 2. Finding the selected vertex v_{ps}
Based on the associated wights of a vertex v_i, namely t_i, T_i, l_i, and the associated weight of the requested data item d_j, namely $N(d_j)$, a particular scheduling algorithm selects a vertex v_{ps} with the highest scheduling priority value (P) among all the vertices in $V(G)$. For instance, EDF selects the vertex v_{ps} with the lowest T_p value, MRF selects a vertex v_{ps} with the highest $N(d_s)$, LTSF selects a vertex v_{ps} with the highest $\frac{t_p}{\frac{l_s}{B}}$ value. The details of v_{ps} selection criteria of each of the scheduling algorithm are shown in Section 4.2.

Step 3. Encoding and broadcasting the encoded packet $\beta(\delta_{max}^{v_{ps}})$
After finding the selected vertex v_{ps}, in this step the heuristic approach finds the maximal clique $\delta_{max}^{v_{ps}}$ which covers v_{ps}. For finding the maximal clique, we use the approach presented in [14]. After finding the maximal clique $\delta_{max}^{v_{ps}}$, the next job is to encode the packet $\beta(\delta_{max}^{v_{ps}})$ which consists all the corresponding data items in $\delta_{max}^{v_{ps}}$. After broadcasting $\beta(\delta_{max}^{v_{ps}})$, the system updates the CR-graph G.

For n number of vertices the Step 1 has $O(n)$ complexity, Step 2 has to search the selected vertex v_{ps} which incurs $O(n)$ complexity, and the Step 3 needs to find the maximal clique for a specific selected vertex v_{ps} from the constructed (in Step 1) CR-graph G, which incurs $O(n^3)$ complexity. So, the total complexity of the proposed heuristic coding based broadcast approach is $O(n)+O(n)+O(n^3) \approx O(n^3)$, which is practical to implement.

4.2. Scheduling Algorithms with Heuristic Coding Implementation

We select a number of representative on-demand scheduling algorithms from each group and present the implementation of their coding versions. In particular, we analyze how the coding versions of different algorithms can fulfil their original scheduling objectives.

4.2.1. Real-Time Algorithms

- EDF [25]: EDF serves the request with the shortest slack time. In other words, it prioritizes the request urgency in scheduling. The network coding version of EDF, needs to find the largest clique δ which contains the vertex v_i with the minimum slack time T_{min} among all the vertices in graph G. In the heuristic coding implementation (denoted by EDF_N), the system first finds the selected vertex v_{ps} with T_{min}. In Algorithm 1, this is done in Step 2. v_{ps} is the vertex with the highest EDF scheduling priority, i.e., $v_{ps} = \{v_{ij} | T_i = max\{\frac{1}{T_1}, \frac{1}{T_2}, \cdots, \frac{1}{T_n}\}, 1 \leq j \leq m\}$. Then Step 3 searches for the maximal clique $\delta_{max}^{v_{ps}}$ to ensure that the encoded packet $\beta(\delta_{max}^{v_{ps}})$ consists the requested data item of the most urgent client.

- SIN [11]: SIN broadcasts the item with the minimum *SIN* value, which is the ratio of the slack time of the most urgent pending request to the number of pending requests for that data item. The network coding version of SIN, needs to find the clique δ in G with the maximum $\frac{|\delta|}{T_{min}}$ value, where T_{min} is the minimum value among all T_i in δ. In the heuristic coding implementation (denoted by SIN_N), the system first finds the selected vertex v_{ps} with the minimum $\frac{T_p}{N(d_s)}$ value, In Step 2 of Algorithm 1, the v_{ps} selection is done by finding the vertex with the maximum $\frac{N(d_s)}{T_p}$ value, i.e., $v_{ps} = \left\{v_{ij} | \frac{N(d_j)}{T_i} = max\left\{\frac{N(d_{r(1)})}{T_1}, \frac{N(d_{r(2)})}{T_2}, \cdots, \frac{N(d_{r(n)})}{T_n}, 1 \leq r, j \leq m\right\}\right\}$. Then Step 3 searches for the maximal clique $\delta_{max}^{v_{ps}}$ to ensure that the encoded packet $\beta(\delta_{max}^{v_{ps}})$ consists the requested data item of the client with the minimum *SIN* value.

4.3. Non-Real-Time Algorithms

- FCFS [26]: FCFS serves requests according to their arrival order. The network coding version of FCFS, needs to find the largest clique δ which contains the vertex v_i with longest waiting time t_{max} among all t_i in G. In the heuristic coding implementation (denoted by FCFS_N), the system first finds the selected vertex v_{ps} with t_{max}. In Step 2 of Algorithm 1, v_{ps} is the vertex with the highest FCFS scheduling priority, i.e., $v_{ps} = \{v_{ij} | t_i = max\{t_1, t_2, \cdots, t_n\}, 1 \leq j \leq m\}$. Then Step 3 searches for the maximal clique $\delta_{max}^{v_{ps}}$ to ensure that the encoded packet $\beta(\delta_{max}^{v_{ps}})$ consists the requested data item of the client with the longest waiting time.

- MRF [27]: MRF broadcasts the data item with the largest number of pending requests. The network coding version of MRF, needs to find the maximum clique δ_{max} in G, where δ_{max} is the maximum clique among all the possible cliques in G. In the heuristic coding implementation (denoted by MRF_N), the system first finds the selected vertex v_{ps} with the maximum $N(d_s)$ value. In Step 2 of Algorithm 1, v_{ps} is the vertex with the highest MRF scheduling priority, i.e., $v_{ps} = \{v_{ij} | N(d_j) =$

$max\{N(d_1), N(d_2), \cdots, N(d_m)\}, 1 \leq i \leq n\}$. Then Step 3 searches for the maximal clique $\delta_{max}^{v_{ps}}$ to ensure that the encoded packet $\beta(\delta_{max}^{v_{ps}})$ consists the requested data item with the largest number of pending requests.

- LWF [27]: LWF broadcasts the data item with the largest total waiting time of all the pending requests for that item. The network coding version of LWF, needs to find the largest clique δ in G with the maximum summed waiting time of the corresponding vertices. In the heuristic coding implementation (denoted by LWF_N), the system first finds the selected vertex v_{ps} with the maximum sum waiting time ($\sum t_i$, $1 \leq i \leq n$) value of data item d_s, denoted as $\sum t(d_s)$. In Step 2 of Algorithm 1, v_{ps} is the vertex with the highest LWF scheduling priority, i.e., $v_{ps} = \{v_{ij} | \sum t(d_j) = max\{\sum t(d_1), \sum t(d_2), \cdots, \sum t(d_m)\}, 1 \leq i \leq n\}$. Then Step 3 searches for the maximal clique $\delta_{max}^{v_{ps}}$ to ensure that the encoded packet $\beta(\delta_{max}^{v_{ps}})$ consists the requested data item with the largest total waiting time of the pending requests.

- R×W [4]: R×W schedules the data item with the maximum R×W value, where R is the number of pending requests for that data item and W is the waiting time of the earliest request for that data item. The network coding version of R×W, needs to find the largest clique δ with the maximum $|\delta| \times t_{max}$ value, where t_{max} is the maximum waiting time value among all t_i in δ. In the heuristic coding implementation (denoted by R×W_N), the system first finds the selected vertex v_{ps} with the maximum $N(d_s) \times t_p$ value. In Step 2 of Algorithm 1, the v_{ps} selection is done by finding the vertex with the maximum $N(d_s) \times t_p$ value, i.e., $v_{ps} = \{v_{ij} | N(d_j) \times t_i = max\{N(d_{r(1)}) \times t_1, N(d_{r(2)}) \times t_2, \cdots, N(d_{r(n)}) \times t_n, 1 \leq r, j \leq m\}\}$. Then Step 3 searches for the maximal clique $\delta_{max}^{v_{ps}}$ to ensure that the encoded packet $\beta(\delta_{max}^{v_{ps}})$ consists the requested data item with the maximum R×W value.

4.4. Stretch Optimal Algorithms

- LTSF [29]: LTSF broadcasts the data item with the largest total current stretch, where stretch is the ratio of the request waiting time to the service time of the data item. The network coding version of LTSF, needs to select the largest clique δ with the maximum $\frac{\sum_{v_i \in \delta} t_i}{max_{v_i \in \delta}\{\frac{l_i}{B}\}}$ value, where $\sum_{v_i \in \delta} t_i$

is the total waiting time of the vertices in δ and $max_{v_i \in \delta}\{\frac{li}{B}\}$ is the service time of the vertices in δ. Note that the service time of the coded packet will be the service time of the largest data item in the clique. In the heuristic coding implementation (denoted by LTSF_N), the system first finds the selected vertex v_{ps} with the largest total current stretch $\left(max\left\{\frac{\Sigma t(d_s)}{\frac{l_p}{B}}\right\}\right)$ for the data item d_s. In Step 2 of Algorithm 1, the v_{ps} selection is done by finding the vertex with the maximum $\left\{\frac{\Sigma t(d_s)}{\frac{l_p}{B}}\right\}$ value, i.e., $v_{ps} = \left\{v_{ij} \mid \left\{\frac{\Sigma t(d_j)}{\frac{l_j}{B}}\right\} = max\left\{\left\{\frac{\Sigma t(d_1)}{\frac{l_1}{B}}\right\}, \left\{\frac{\Sigma t(d_2)}{\frac{l_2}{B}}\right\}, \cdots, \left\{\frac{\Sigma t(d_m)}{\frac{l_m}{B}}\right\}, 1 \le i \le n\right\}\right\}$. Then Step 3 searches for the maximal clique $\delta_{max}^{v_{ps}}$ to ensure that the encoded packet $\beta(\delta_{max}^{v_{ps}})$ consists the requested data item with the largest total current stretch.

- STOBS [28]: STOBS broadcasts the data item with the largest $\frac{R \times W}{S}$ value, where R, W and S represent the number of pending requests, waiting time of the earliest pending request and the item size, respectively. The network coding version of STOBS, needs to find the largest clique δ with the maximum $\frac{|\delta| * max_{v_i \in \delta} t_i}{max_{v_i \in \delta} l_i}$ value, where $max_{v_i \in \delta} t_i$ and $max_{v_i \in \delta} l_i$ are respectively the maximum waiting time and the maximum item size among all the vertices in δ. Note that $max_{v_i \in \delta} l_i$ also denotes the size of the encoded packet. In the heuristic coding implementation (denoted by STOBS_N), the system first finds the selected vertex v_{ps} with the maximum $\frac{N(d_s) \times t_p}{l_s}$ value. In Step 2 of Algorithm 1, the v_{ps} selection is done by finding the vertex with the maximum $\frac{N(d_s) \times t_p}{l_s}$ value, i.e., $v_{ps} = \left\{v_{ij} \mid \frac{N(d_j) \times t_i}{l_j} = max\left\{\frac{N(d_{r(1)}) \times t_1}{l_{r(1)}}, \frac{N(d_{r(2)}) \times t_2}{l_{r(2)}}, \cdots, \frac{N(d_{r(n)}) \times t_n}{l_{r(n)}}, 1 \le r, j \le m\right\}\right\}$. Then Step 3 searches for the maximal clique $\delta_{max}^{v_{ps}}$ to ensure that the encoded packet $\beta(\delta_{max}^{v_{ps}})$ consists the requested data item with the largest $\frac{R \times W}{S}$ value.

4.5. Running Example

We give a running example to better illustrate the transformation of algorithms into their network coding versions based on the proposed heuristic coding based approach and analyze their potential performance improvement. As shown in

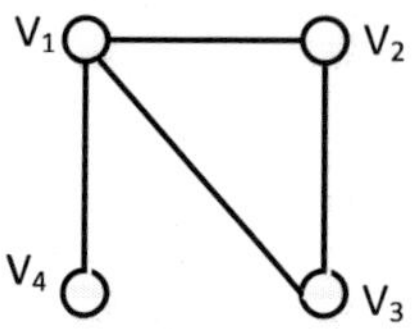

Vertex	Request	Cache	Waiting time	Slack time
v_1	$q_1 \rightarrow d_1$	$\{d_2, d_3\}$	$t_1=2$	$T_1=2$
v_2	$q_2 \rightarrow d_1$	$\{d_2\}$	$t_2=4$	$T_2=3$
v_3	$q_3 \rightarrow d_2$	$\{d_1\}$	$t_3=5$	$T_3=1$
v_4	$q_4 \rightarrow d_3$	$\{d_1\}$	$t_4=7$	$T_4=2$
Broadcast bandwidth = 1.0				

Figure 2. A running example.

Figure 2, suppose there are three data items d_1, d_2, d_3. Client c_1 has cached d_2 and d_3. Similarly, clients c_2, c_3 and c_4 have cached d_2, d_1 and d_1, respectively. Consider four requests q_1, q_2, q_3 and q_4 asking for d_1, d_1, d_2 and d_3, respectively, and q_i is submitted by c_i ($1 \leq i \leq 4$). The current waiting time of q_1, q_2, q_3 and q_4 are $t_1 = 2$, $t_2 = 4$, $t_3 = 5$ and $t_4 = 7$, respectively. Assume that the bandwidth is $B = 1\,{}^{\text{unit}}/_{\text{tick}}$. For the real-time and the non-real-time scheduling algorithms, consider that the item size is 1 unit, and hence the server broadcasts one data item at every broadcast tick. For the real-time algorithms (i.e., EDF and SIN), we assume that the current slack time of q_1, q_2, q_3 and q_4 are $T_1 = 2$, $T_2 = 3$, $T_3 = 1$ and $T_4 = 2$, respectively. For the stretch optimal algorithms (i.e., LTSF and STOBS), we assume that the sizes of d_1, d_2, d_3 are 3 units, 2 units and 4 units, respectively. Accordingly, we have $l_1 = 3$, $l_2 = 3$, $l_3 = 2$ and $l_4 = 4$.

According to Definition 1, a graph G is constructed as shown in Figure 2, where vertices v_1, v_2, v_3 and v_4 correspond to requests q_1, q_2, q_3 and q_4, respectively. Since both q_1 and q_2 request the same data item d_1, there is an edge between v_1 and v_2. Since c_1 caches data item d_2, which is requested by c_3 (i.e., q_3), and c_3 caches d_1 which is requested by c_1 (i.e., q_1), there is an edge between v_1 and v_3. Similarly, other edges are constructed following the same rules. In the following, we analyze and compare the performance of each algorithm with respect to their non-coding and coding versions.

As EDF and SIN correspond to the real-time model, we show the comparative performance difference between coding and non-coding versions of them in number of deadline missed requests.

- EDF: The broadcast sequence in the non-coding version is $\{d_2, d_1, d_3\}$, since the order of requests according to the request urgency is q_3, q_1, q_4, q_2. The deadline miss ratio is 25%, because after broadcasting d_2 and d_1, q_4 misses its deadline. In the heuristic coding implementation, v_3 is the *selected vertex* as v_3 corresponds to the most urgent request q_3, i.e., T_3

equals 1. Hence in the coding version, the maximal clique is $\{v_1, v_2, v_3\}$, which covers the selected vertex v_3. Thus, the broadcast sequence is $\{d_1 \oplus d_2, d_3\}$, and no request misses its deadline.

- SIN: The broadcast sequence of the non-coding SIN version is $\{d_2, d_1, d_3\}$. The deadline miss ratio is 25%. In the heuristic coding implementation, v_3 is the *selected vertex* and the maximal clique is $\{v_1, v_2, v_3\}$, covers the selected vertex v_3. Hence, the broadcast sequence is $\{d_1 \oplus d_2, d_3\}$, and no request misses its deadline.

As FCFS, MRF, LWF and R×W correspond to the non-real-time model, we show the comparative performance difference between coding and non-coding versions of them in access time.

- FCFS: The broadcast sequence of the non-coding version is $\{d_3, d_2, d_1\}$, and the total access time is $8(q_4) + 7(q_3) + 7(q_2) + 5(q_1) = 27$. In the heuristic coding implementation, v_4 is the *selected vertex* and the maximal clique is $\{v_1, v_4\}$. Thus, the broadcast sequence is $\{d_1 \oplus d_3, d_1 \oplus d_2\}$, and the total access time is $3(q_1) + 8(q_4) + 6(q_2) + 7(q_3) = 24$.

- MRF: The broadcast sequence of the non-coding version is $\{d_1, d_2, d_3\}$, and the total access time is $3(q_1) + 5(q_2) + 7(q_3) + 10(q_4) = 25$. In the heuristic coding implementation, either v_1 or v_2 can be the *selected vertex* and the maximal clique is $\{v_1, v_2, v_3\}$. Hence, the broadcast sequence is $\{d_1 \oplus d_2, d_3\}$, and the total access time is $3(q_1) + 5(q_2) + 6(q_3) + 9(q_4) = 23$.

- LWF: The broadcast sequence of the non-coding version is $\{d_3, d_1, d_2\}$, and the total access time is $8(q_4) + 4(q_1) + 6(q_2) + 8(q_3) = 26$. In the heuristic coding implementation, v_4 is the *selected vertex* and the maximal clique is $\{v_1, v_4\}$. Thus, the broadcast sequence is $\{d_1 \oplus d_3, d_1 \oplus d_2\}$, and the total access time is $3(q_1) + 8(q_4) + 6(q_2) + 7(q_3) = 24$.

- R× W: The broadcast sequence of the non-coding version is $\{d_1, d_3, d_2\}$ and the total access time is $3(q_1) + 5(q_2) + 9(q_4) + 8(q_3) = 25$. In the heuristic coding implementation, v_2 is the *selected vertex* and the maximal clique is $\{v_1, v_2, v_3\}$. Thus, the broadcast sequence is $\{d_1 \oplus d_2, d_3\}$, and the total access time is $3(q_1) + 5(q_2) + 6(q_3) + 9(q_4) = 23$.

As LTSF and STOBS correspond to the stretch optimal model, we show the comparative performance difference between coding and non-coding versions of them in stretch. The sizes of d_1, d_2, d_3 are set to 3, 2, 4, respectively.

- LTSF: The broadcast sequence of the non-coding version is $\{d_2, d_1, d_3\}$, and the total stretch is $\frac{5+2}{2}(q_3) + \frac{2+3+2}{3}(q_1) + \frac{4+3+2}{3}(q_2) + \frac{7+4+2+3}{4}(q_4) = 12.8$. In the heuristic coding implementation, v_3 is the *selected vertex*. Hence in the coding version, the maximal clique is $\{v_1, v_2, v_3\}$. Thus, the broadcast sequence is $\{d_1 \oplus d_2, d_3\}$, and the service order is q_1, q_2, q_3, q_4. Note that the size of the encoded data item equals the size of the largest data item being encoded. Hence, the total stretch is $\frac{2+3}{3}(q_1) + \frac{4+3}{3}(q_2) + \frac{5+3}{2}(q_3) + \frac{7+4+3}{4}(q_4) = 11.5$.

- STOBS: The broadcast sequence of the non-coding is $\{d_1, d_2, d_3\}$, and the total stretch is $\frac{2+3}{3}(q_1) + \frac{4+3}{3}(q_2) + \frac{5+2+3}{2}(q_3) + \frac{7+4+2+3}{4}(q_4) = 13.0$. In the heuristic coding implementation, v_2 is the *selected vertex* and the maximal clique is $\{v_1, v_2, v_3\}$. Thus, the broadcast sequence is $\{d_1 \oplus d_2, d_3\}$, and the total stretch is $\frac{2+3}{3}(q_1) + \frac{4+3}{3}(q_2) + \frac{5+3}{2}(q_3) + \frac{7+4+3}{4}(q_4) = 11.5$.

Table 3. Simulation parameters

Parameter	Default	Range	Description
ClientNum	400	100∼600	Number of clients
THINKTIME	0.01	—	Request generation interval
m	1000	—	Size of the database
SIZEMIN, *SIZEMAX*	1, 30	—, 10∼60	Min. and Max. data item size
CacheSize	60	30∼180	Client cache size
θ	0.4	0.0∼1.0	Zipf distribution parameter
μ^-, μ^+	120, 200	80∼160, 160∼240	Min. and Max. laxity

5. Simulation Model

5.1. Setup

The simulation model is built based on the system architecture illustrated in Figure 1. It is implemented by CSIM19 [50]. Unless stated otherwise, the simulation is conducted under the default setting. The primary parameters are summarized in Table 3.

It is a closed system model, where a client submits a new request only when the previous one is satisfied or misses its deadline [49]. The $THINKTIME$ refers to the time interval of generating new requests, which follows the Exponential distribution. Each request asks for one data item. The data access pattern follows the commonly used Zipf distribution [51], where the skewness is controlled by the parameter θ. Specifically, when $\theta = 0$, it follows the Uniform distribution. With an increase value of θ, the data access pattern is getting more skewed. Based on a certain scheduling algorithm, the server encodes the requested data items and disseminate them via the broadcast channel. Clients monitor the broadcast channel and store the received data items into the local cache. Each client has the same cache size and the LRU cache replacement policy is adopted.

In real-time environments, the relative deadline of a request q_i (RD_i) is computed by:

$$RD_i = (1 + uniform(\mu^-, \mu^+)) * T_i^{serv}$$

Accordingly, The deadline of q_i (Dl_i) is computed by:

$$Dl_i = AT_i + RD_i$$

where AT_i is the arrival time of q_i. μ^- *and* μ^+ represent the minimum and the maximum laxity, respectively. T_i^{serv} represents the service time of q_i. A request is feasible for serving only if its slack time is longer than the service time. Infeasible requests will be removed from the service queue.

The channel bandwidth is 1 unit/tick. That is, the server disseminates one unit-sized data item in each broadcast tick. The scheduling is non-preemptive, which means that the transmission of the current data item cannot be interrupted [29]. For the real-time and the non-real-time scheduling, the item size is 1 unit. For the stretch optimal scheduling, the sizes of data items are generated using random distribution [52]:

$$DataItemSize[i] = SIZEMIN + \lfloor random(0.0, 1.0) \times (SIZEMAX - SIZEMIN + 1) \rfloor$$

where $SIZEMIN$ and $SIZEMAX$ are the minimum and the maximum data item size in the database, respectively, and $i = 1, 2, \cdots, m$.

5.2. Metrics

We design the following metrics to give an intensive performance analysis.

- Deadline Miss Ratio (DMR): It is the ratio of the number of requests missing deadlines to the total number of submitted requests. It measures the system capability on satisfying requests before their deadlines expire. The primary objective of a real-time scheduling algorithm is to minimize the DMR.
- Average Response Time (ART): It is the duration from the instant when a request is submitted to the time when the corresponding data item is received. It measures the responsiveness of the system. The primary objective of a non-real-time scheduling algorithm is to minimize the ART.
- Average Stretch (AS): It is the ratio of the response time to the service time of the corresponding data item. It is a widely used metric in heterogeneous database environments. The primary objective of a stretch optimal scheduling algorithm is to minimize the AS.
- Average Encode Length (AEL): It is the average number of data items which are selected to construct an encoded packet. It measures the coding flexibility and the power of exploiting coding opportunity of a scheduling algorithm. If the AEL is large, more clients can decode their required data items from an encoded packet.
- Average Broadcast Productivity (ABP): It is the average number of requests satisfied by each broadcast. It measures the productivity of a broadcasted packet/item. A large ABP implies the efficient bandwidth utilization.

6. PERFORMANCE EVALUATION

In this section, we evaluate the performance of both the non-coding and the coding versions of real-time, non-real-time and stretch optimal scheduling algorithms in a variety of circumstances. The results were obtained when the system was in a steady state and the simulations continued until a confidence interval of 0.95 with half-widths of less than 5% about the mean was received.

6.1. Simulation Results in Real-Time Models

Figures 3$\sim$ 6 show the comparative performance graphs of the real-time algorithms for both the non-coding and the coding versions.

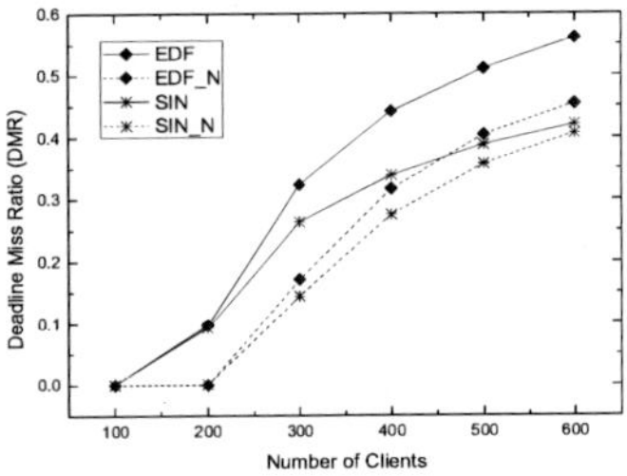

(a) Deadline Miss Ratio (DMR).

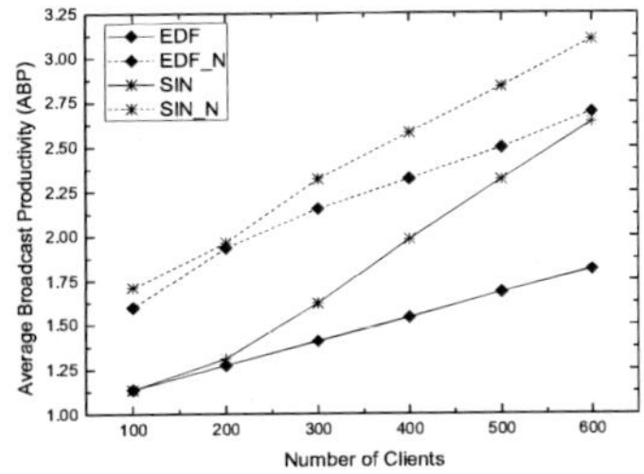

(b) Average Broadcast Productivity (ABP).

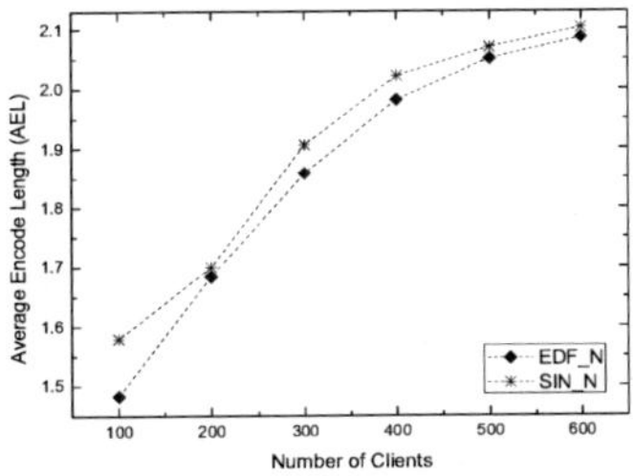

(c) Average Encode Length (AEL).

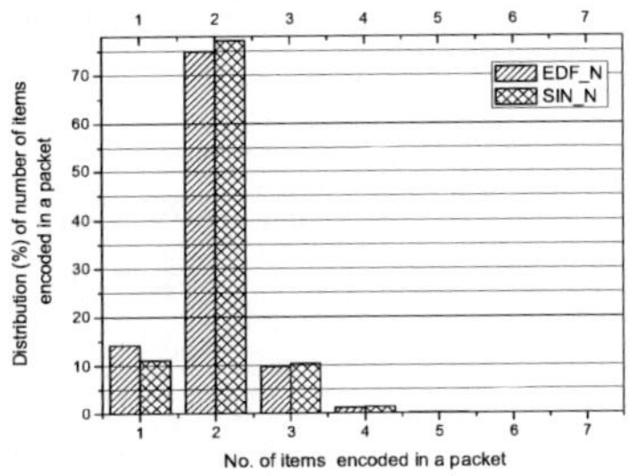

(d) Distribution of number of data items encoded in a packet under default settings.

Figure 3. Impact of workload on the real-time model.

6.1.1. Impact of Workload

Figure 3 shows the impact of workload under different number of clients. When the number of clients increases, the system workload becomes heavier. As a result, more requests miss their deadlines and the algorithms' performances decline (Figure 3(a)). In the non-coding versions of the real-time algorithms, SIN outperforms EDF, because EDF only considers the request urgency. On the contrary, SIN considers both the request urgency and the item popularity, and hence SIN satisfies more requests than EDF (Figure 3(b)). The performance difference between EDF and SIN becomes more significant with an increase workload. The observations are consistent with the results reported in previous studies [11, 53].

The network coding versions of EDF and SIN, namely, EDF_N and SIN_N, outperform their respective non-coding versions notably in terms of achieving the lower deadline miss ratio (DMR). For instance, with the moderate system workload (default setting) this performance improvement is 28% and 19%, respectively for EDF_N and SIN_N. The main reason of this performance improvement is that the coding version achieves a higher average broadcast productivity (ABP) than the corresponding non-coding version (Figure 3(b)). This result shows that more requests can be satisfied by a single broadcast with network coding.

Figure 3(b) shows that with an increase number of clients, the ABP of both the coding and non-coding version increases. With the growing number of requests, the popularity of data items increases. Therefore, for the non-coding version, in spite of a heavier system workload, the ABP increases. For the coding version, more requests provide more coding opportunities (Figure 3(c)), thereby yielding a higher ABP (Figure 3(b)). Note that the amount of performance improvement of using network coding for EDF is larger than SIN and the amount of performance improvement of SIN_N over SIN diminishes when the number of clients is large. This is because network coding indirectly endows the item popularity attribute to EDF when it chooses the largest clique that contains the data item with the earliest deadline. So, the performance improvement of EDF_N is more significant and it increases with the item popularity. However, since SIN has already considered the item popularity, the performance gain due to network coding in SIN_N is less significant. On the whole, SIN_N achieves the highest ABP and the lowest DMR (Figure 3(a)).

Figure 3(c) shows that SIN_N consistently achieves a higher average encode length (AEL) than EDF_N. The result shown in Figure 3(d) explores the reason. In particular, it shows the percentage distribution of number of items in an encoded packet under the default setting. Most of the encoded packets consist of two items, while the rest packets consist of one or three items. Specifically, SIN_N gives a higher chance of encoding more than one data items in a packet than EDF_N. For instance, 77.2% of the packets encoded by SIN_N consist of two data items, while 74.8% of the packets encoded by EDF_N consist of two data items. On the other hand, the number of items which are not encoded in SIN_N is less than that of EDF_N. Specifically, 11% of the items are not encoded in SIN_N, while 14.2% of the items are not encoded in EDF_N. In general, a higher ABP and AEL can be achieved if there are more encoded packets which consists of more data items. This has been demonstrated in Figure 3(b) and Figure 3(c), respectively.

6.1.2. Impact of Skewness Parameter (θ)

Figure 4 compares the algorithm performance under different data access patterns. When $\theta = 0$, all the data items in the database have the same access probability. With an increase of θ, the data access pattern is getting skewed. Previous studies [4, 11, 23] have demonstrated that algorithms will perform better when the data access pattern is getting skewed. This is because each broadcast data item has the potential to satisfy more requests. Accordingly, the ABP of each algorithm increases (Figure 4(b)). For the coding version, when the data access pattern is skewed, for those data items which are requested more frequently, they also have higher chance to be cached by the clients. This provides more coding opportunities and leads to a larger AEL. For instance, when $\theta = 1.0$, the AEL of SIN_N is close to 3.0 (Figure 4(c)). This means that by broadcasting an encoded packet, three data items can be disseminated simultaneously. Inevitably, the ABP of the coding version of an algorithm is much higher than its non-coding version. Figure 4(a) shows that the coding versions of algorithms outperform the non-coding versions across the whole range in terms of achieving lower DMRs, and SIN_N has the best performance among all the algorithms.

6.1.3. Impact of Cache Size

Figure 5 depicts the impact of the cache size on the coding versions of the real-time algorithms. Since the non-coding version of an algorithm does not utilize cache contents in scheduling, the cache size does not have any impact on non-coding versions. In the coding version, more data items can be stored with an increase of cache size, which increases the probability that one client's cached data item happened to be another client's requested data item. Therefore, when the cache size is large, it is likely to produce large cliques, which is the key to exploit the coding flexibility and increase the AEL. Figure 5(c) shows the AELs of EDF_N and SIN_N. They both achieve higher AEL with an increase of cache sizes. Accordingly, the ABPs of the corresponding algorithms also increase (Figure 5(b)). Since a higher ABP indicates that more requests are served in a broadcast tick, the DMRs of the coding versions reduce with the increase of cache size as shown in Figure 5(a).

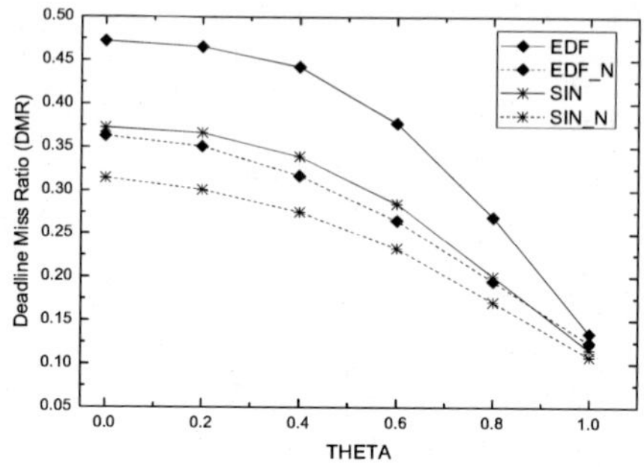

(a) Deadline Miss Ratio (DMR)

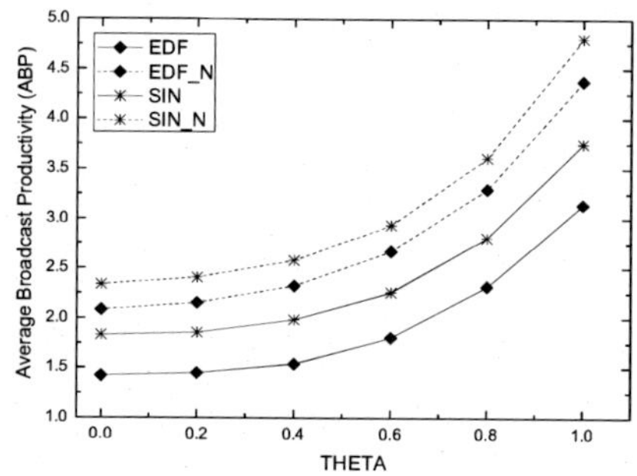

(b) Average Broadcast Productivity (ABP)

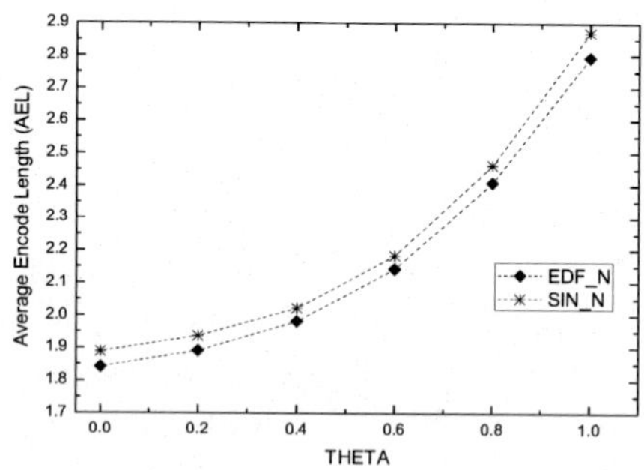

(c) Average Encode Length (AEL)

Figure 4. Impact of skewness parameter (θ) on the real-time model.

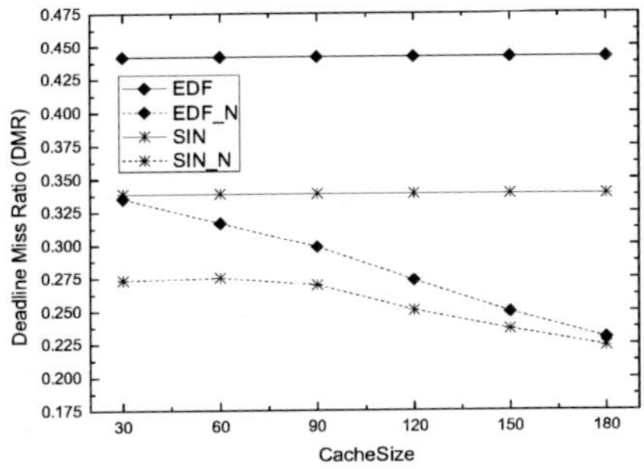

(a) Deadline Miss Ratio (DMR).

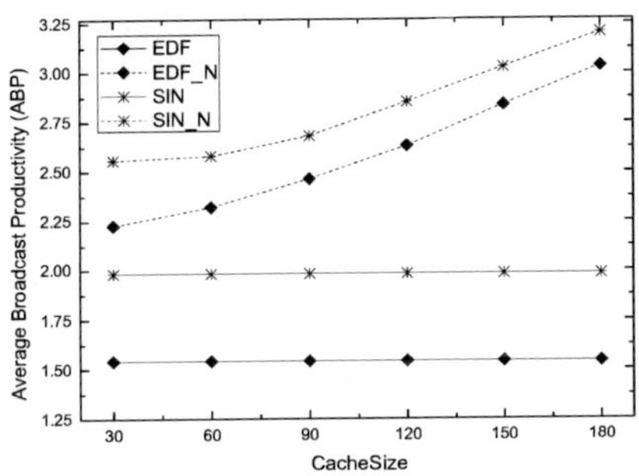

(b) Average Broadcast Productivity (ABP).

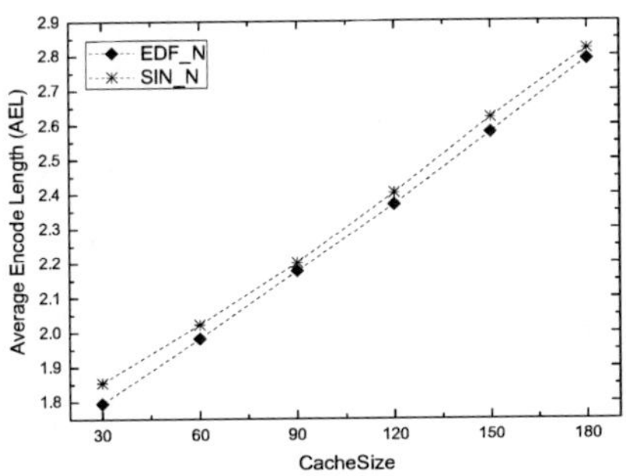

(c) Average Encode Length (AEL).

Figure 5. Impact of cache size on the real-time model.

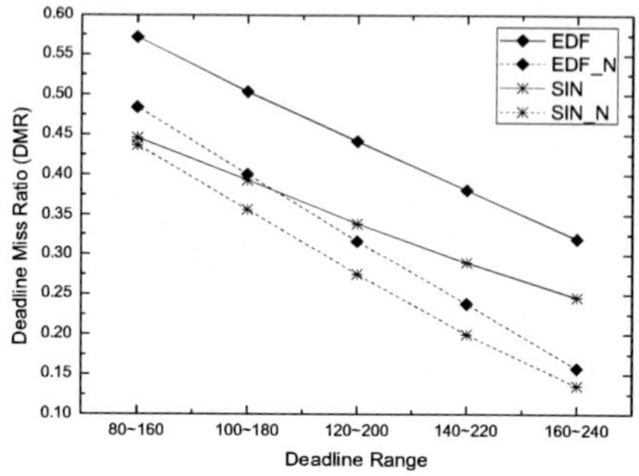

(a) Deadline Miss Ratio (DMR).

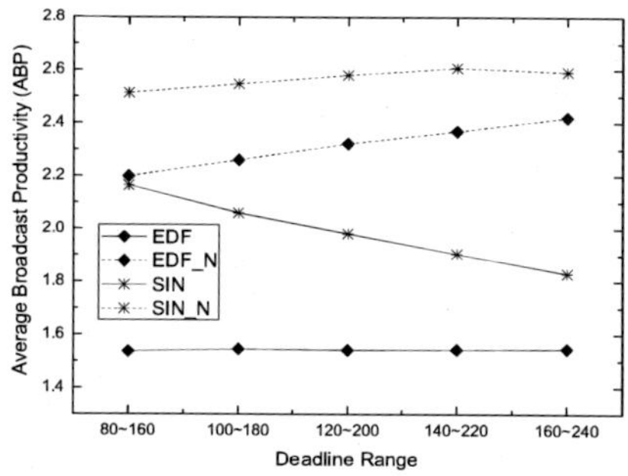

(b) Average Broadcast Productivity (ABP).

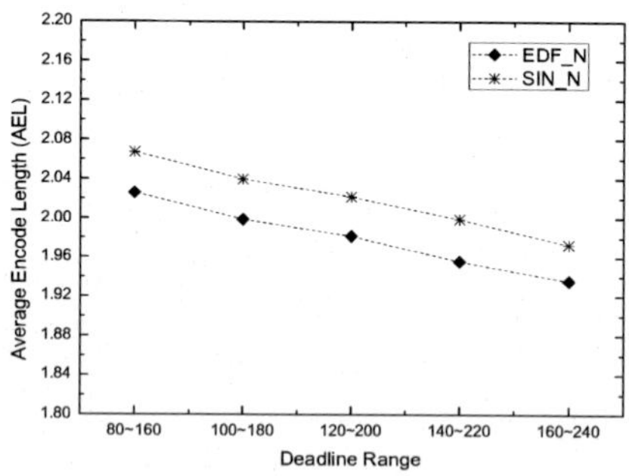

(c) Average Encode Length (AEL).

Figure 6. Impact of deadline range on the real-time model.

6.1.4. Impact of Deadline Range

Recalling that the relative deadline of requests is uniformly selected from the range $\mu^- \sim \mu^+$. A larger value of the relative deadline gives a looser request deadline. Consequently, the DMRs of all the algorithms (Figure 6(a)) drop with an increase mean value of $\mu^- \sim \mu^+$. Evidently, SIN_N outperforms others.

In the coding versions, when the deadline becomes loose, the system has more opportunities to encode and disseminate data items to serve clients. As shown in Figure 6(a) and Figure 6(b), for the coding versions, the DMRs drop and the ABPs increase slightly. However, a loose request deadline does not have much impact on the AELs (< 0.1) of EDF_N and SIN_N (Figure 6(c)). In the non-coding versions, when deadline range increases, the ABP of EDF does not change, whereas SIN shows a downward trend. This is because EDF only considers the request urgency, but SIN considers both the request urgency and the item popularity. As data items with less popularity may have better chance to be disseminated when the request deadline is getting looser, the ABP of SIN drops.

6.2. Simulation Results in Non-Real-Time Models

Figures 7$\sim$ 9 show the comparative performance graphs of the non-real-time algorithms for both the non-coding and the coding versions.

6.2.1. Impact of Skewness Parameter (θ)

Figure 7 shows the impact of the data access pattern θ on the non-real-time algorithms. Evidently, both the non-coding and the coding versions of the algorithms have better performance in terms of minimizing the average response time (ART) with an increase value of θ. This is due to the same reasons as analyzed in Section 6.1.2. In addition, the coding version of an algorithm achieves a much lower response time (Figure 7(b)) than its non-coding version (Figure 7(a)). On an average, irrespective of the skewness of the data access pattern, the coding version of each algorithm has at least 40% less ART than the non-coding version. The relative performance positions of the algorithms in the coding group are identical to the non-coding group. MRF performs the best when the data access pattern is less skewed. When the data access pattern is skewed, MRF becomes the worst because unpopular data items may suffer from starvation, which causes the large ART. The result is consistent with the findings in

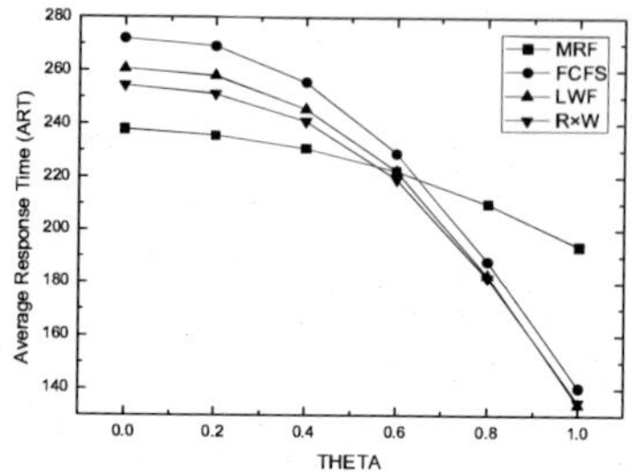

(a) Average Response Time (ART) of the non-coding version.

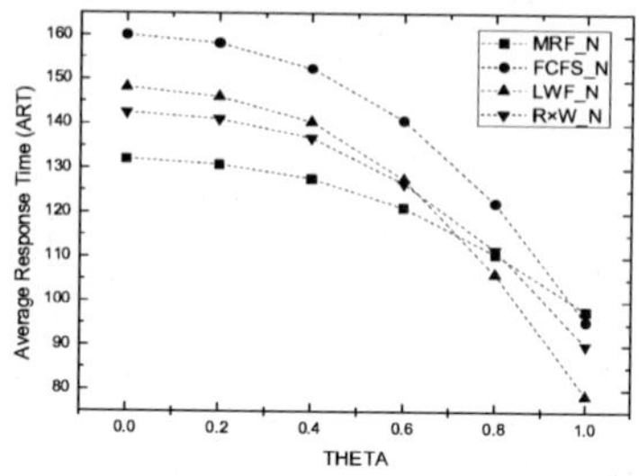

(b) Average Response Time (ART) of the coding version.

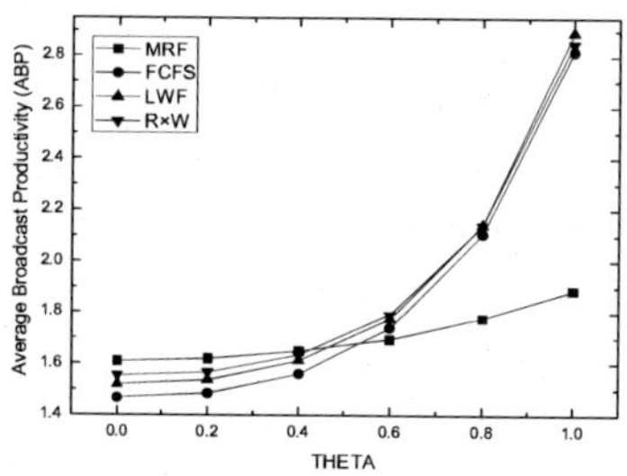

(c) Average Broadcast Productivity (ABP) of the non-coding version.

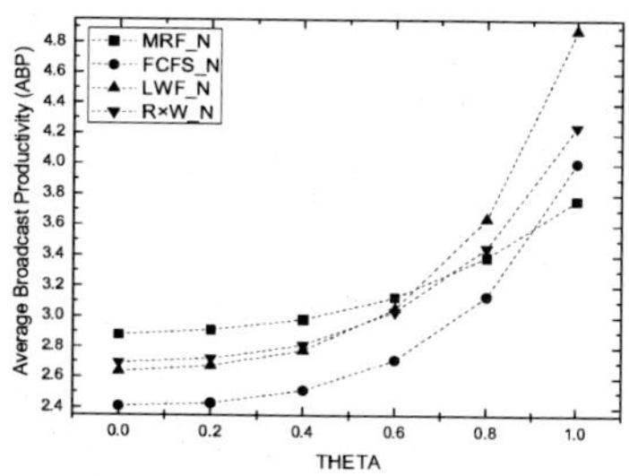

(d) Average Broadcast Productivity (ABP) of the coding version.

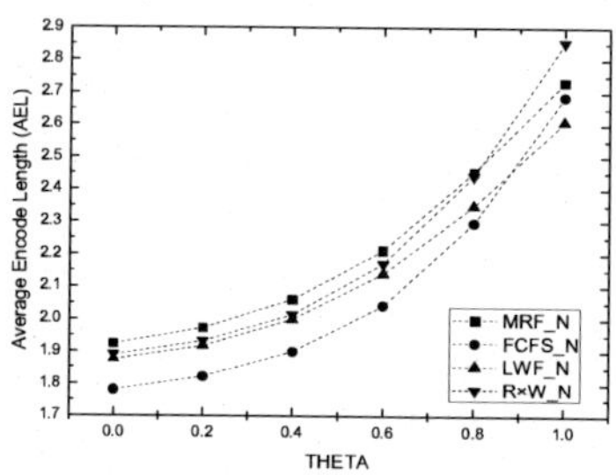

(e) Average Encode Length (AEL) of the coding version.

Figure 7. Impact of skewness parameter (θ) on the non-real-time model.

previous works [13, 35, 54]. MRF_N shows similar trend in Figure 7(b). FCFS which only considers the request waiting time, performs the worst in most cases. Similarly, FCFS_N has the worst performance among the coding versions. R× W and LWF have similar performance, because both the algorithms consider the item popularity and the request waiting time in scheduling. In the coding version, LWF_N significantly outperforms other algorithms in a highly skewed data access pattern environment. In the non-coding version, excepting MRF, the ABPs of all the algorithms increase steadily when the data access pattern is getting skewed (Figure 7(c)). This is because MRF solely considers item popularity in scheduling, resulting in very low ABP for unpopular data items. Similar results are observed in the coding versions of the algorithms (Figure 7(d)). The AELs increase when the data access pattern is getting skewed (Figure 7(e)). These results further explain the relative performances of algorithms as shown in Figures 7(a) and 7(b).

6.2.2. Impact of Workload

Figure 8 shows the impact of workload on the non-real-time scheduling algorithms. As expected, when the system workload is getting higher, the ARTs of both the non-coding and the coding versions increase (Figures 8(a) and 8(b)). MRF and MRF_N have the best performance in the non-coding and the coding versions, respectively. Note that $\theta = 0.4$ in the default setting, where MRF has been shown with satisfactory performance because the data access pattern is not quite skewed. FCFS and FCFS_N perform the worst in the respective groups. R×W and R×W_N have the moderate performance in the non-coding and the coding versions, respectively. LWF and LWF_N rank next to R×W and R×W_N, respectively. Note that the coding version of an algorithm significantly reduces the access time than its non-coding version under different workloads. It demonstrates the advantages of network coding in terms of adapting to the change of system workloads. With an increase of clients, data items will be requested more frequently. This helps to improve the broadcast productivity. Therefore, the ABPs of the non-coding versions increase (Figure 8(c)). Moreover, this helps to improve the coding flexibility, and hence the AELs of the coding versions increase (Figure 8(e)). As both the reasons positively impact on the broadcast productivity of the coding versions, when the number of clients is getting higher, the ABPs of the coding versions increase much more significantly than that of the non-coding versions (Figure 8(d)).

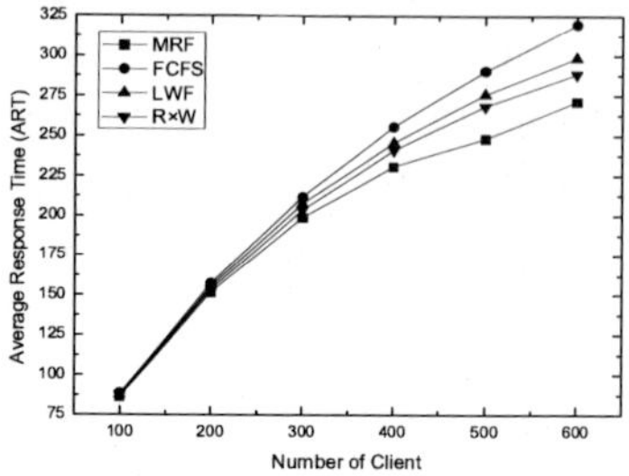

(a) Average Response Time (ART) of the non-coding version.

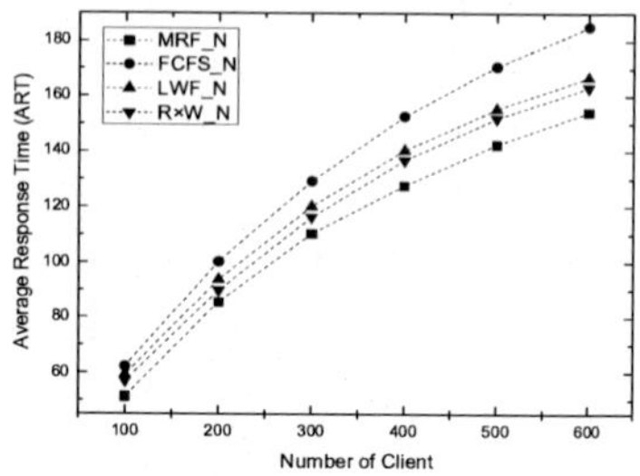

(b) Average Response Time (ART) of the coding version.

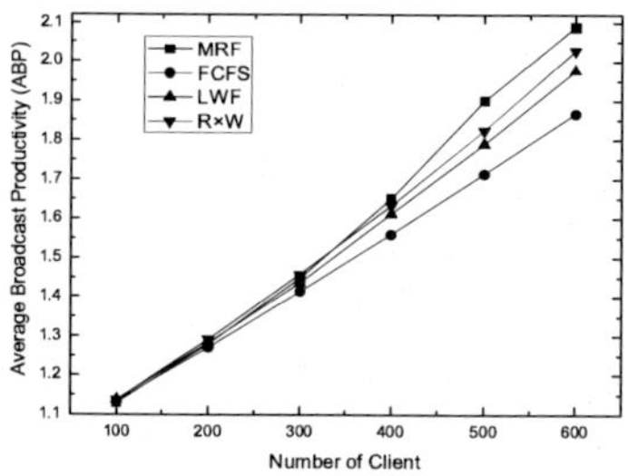

(c) Average Broadcast Productivity (ABP) of the non-coding version.

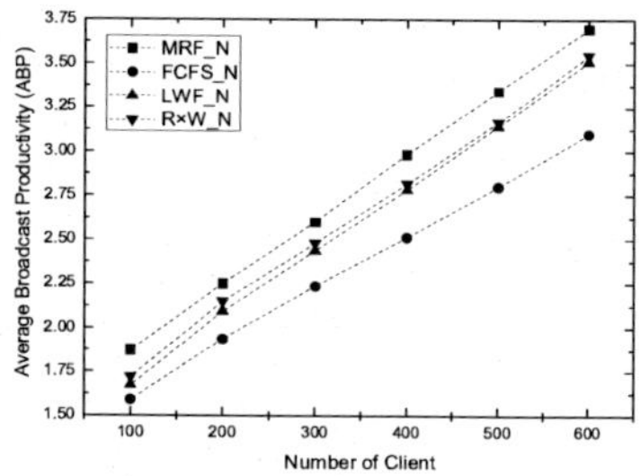

(d) Average Broadcast Productivity (ABP) of the coding version.

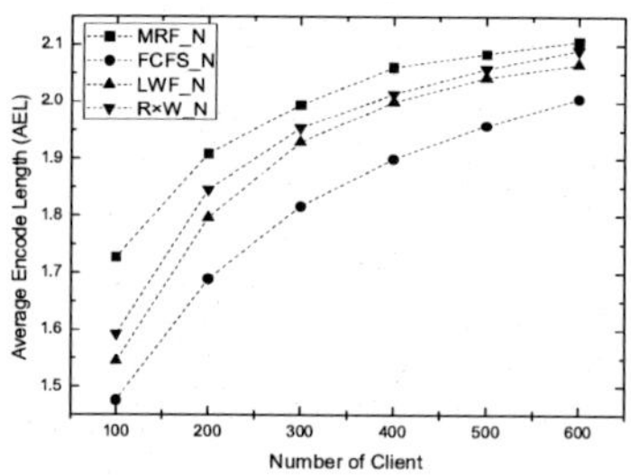

(e) Average Encode Length (AEL) of the coding version.

Figure 8. Impact of workload on the non-real-time model.

6.2.3. Impact of Cache Size

Figure 9 shows the impact of cache sizes on the non-real-time algorithms. When the cache size increases, the coding versions achieve better performance. As discussed in Section 6.1.3, a large cache size helps the coding versions to improve the AELs (Figure 9(b)), and consequently, the ABPs also increase (Figure 9(c)). Therefore, the performance of a coding version improves in terms of achieving the lower ART with an increase of cache size, which is shown in Figure 9(a). Note that even with small cache sizes, network coding may improve the system performance significantly. For instance, as shown in Figure 9(a), when the cache size is small (e.g., 30), the coding versions have much higher ARTs than their non-coding versions.

6.3. Simulation Results in Stretch Optimal Models

Figures 10 $\sim$ 15 show the comparative performance graphs of the stretch optimal scheduling algorithms for both the non-coding and the coding versions. In this set of simulations, the sizes of data items are various. Specifically, the sizes of data items are randomly generated between 1 and 30. Moreover, for comparison purposes, we maintain a constant system workload with various data sizes by changing the $THINKTIME$ accordingly.

6.3.1. Impact of Workload

Figure 10 depicts the behavior of the stretch optimal algorithms under different system workloads. The coding versions show a significant better performance than their non-coding versions in terms of minimizing the ART (Figure 10(a)). With respect to the average stretch (AS), a coding version also outperforms the corresponding non-coding version (Figure 10(b)). Both in the non-coding and the coding versions, LTSF and STOBS have similar performance. It is worth mentioning that, STOBS schedules the data item with the maximum $\frac{R \times W}{S}$ value. In contrast, LTSF schedules the data item with the longest total current stretch, which is the total stretches of all the pending requests of that data item. This implies that LTSF also considers both the data item popularity and the request waiting time. Given the broadcast bandwidth of 1 unit/tick, the time to broadcast a data item (i.e., the number of broadcast ticks) equals the number of units of the data item (i.e., the item size). On a whole, the attributes of LTSF and STOBS are very similar, which result in their close performance. For the im-

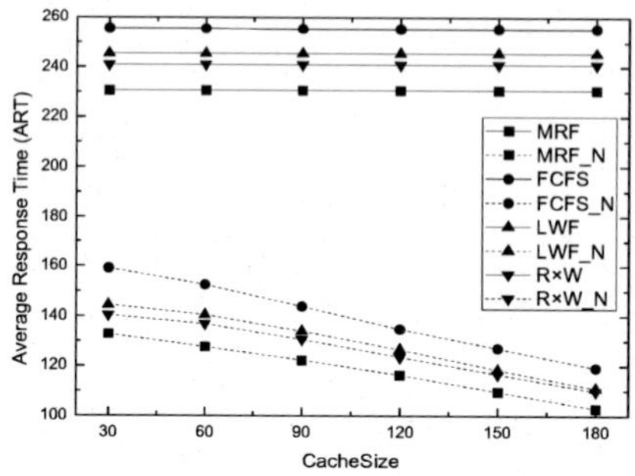

(a) Average Response Time (ART).

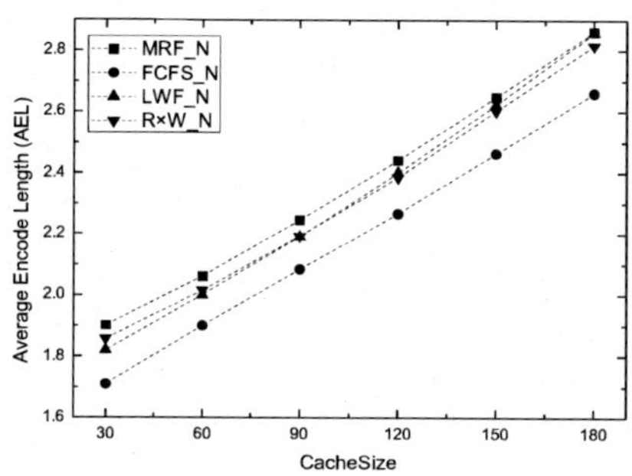

(b) Average Encode Length (AEL).

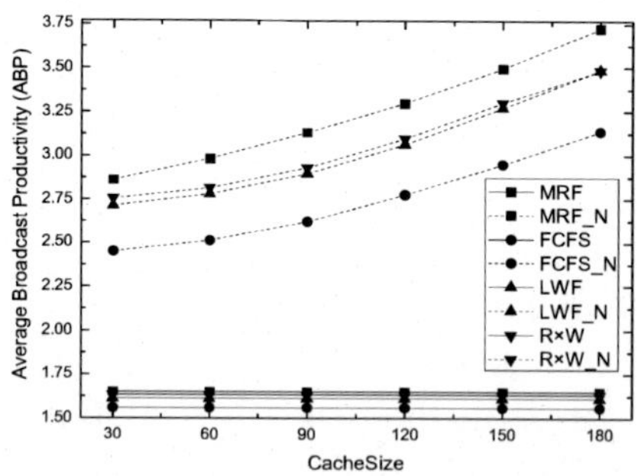

(c) Average Broadcast Productivity (ABP).

Figure 9. Impact of cache size on the non-real-time model.

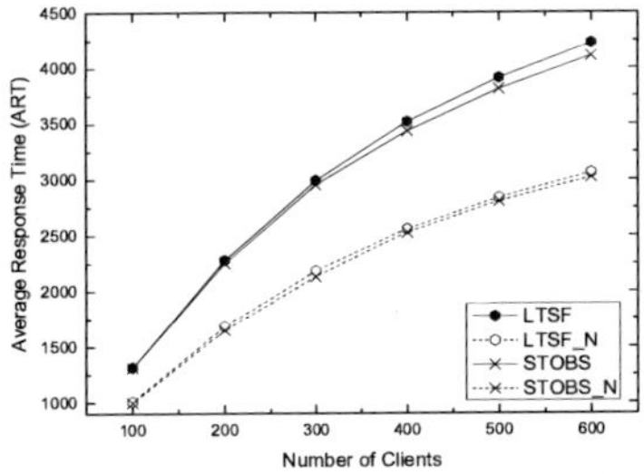

(a) Average Response Time (ART).

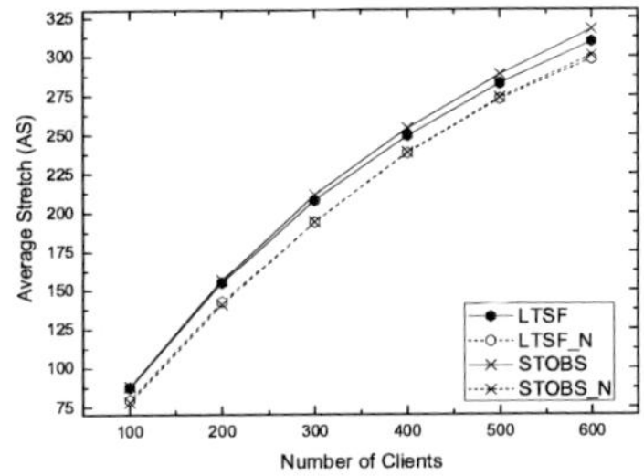

(b) Average Stretch (AS).

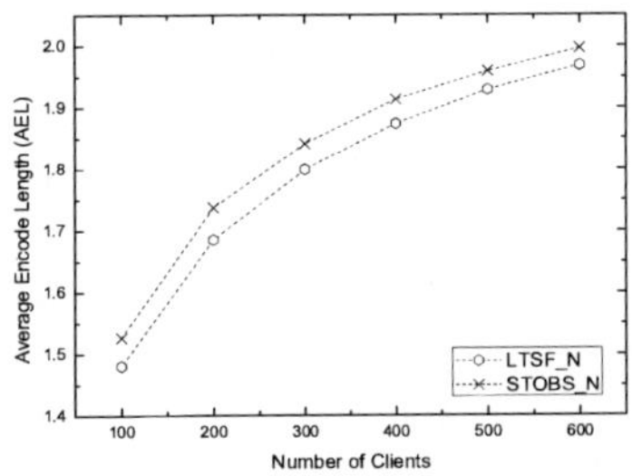

(c) Average Encode Length (AEL).

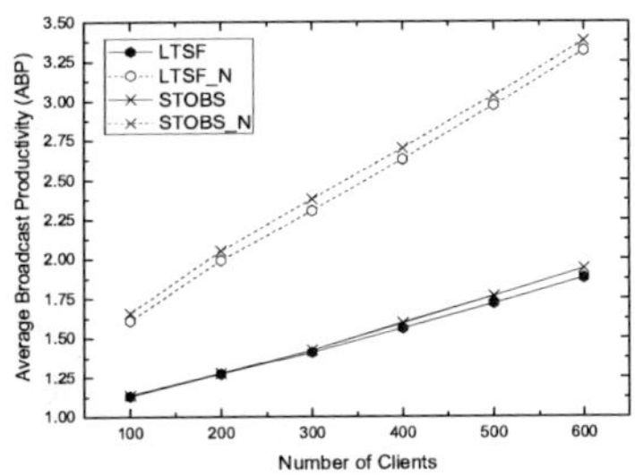

(d) Average Broadcast Productivity (ABP).

Figure 10. Impact of workload on the stretch optimal model.

plementation of their coding versions, STOBS_N computes $\frac{N(d_s)\times t_p}{l_p}$ for finding the *selected vertex* v_{ps}, which is the ratio of the data item popularity multiply the maximum waiting time to the data item size. On the other hand, LTSF_N computes $\left(max \left\{ \frac{\sum t_i}{\frac{l_p}{B}} \right\} \right)$ for finding the *selected vertex* v_{ps}, which is the ratio of the sum of the waiting time of requests to the item size of the corresponding data item. Likewise, the attributes of STOBS_N and LTSF_N are similar, which result in their close performance. Moreover, with an increase of clients, the AEL of a coding version increases (Figure 10(c)), and the ABPs of both versions increase. Note that the ABP of a coding version is much higher than its non-coding version (Figure 10(d)).

6.3.2. Impact of Skewness Parameter (θ)

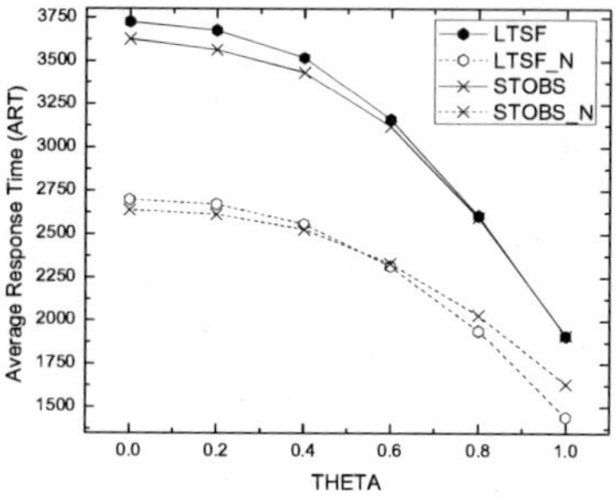

(a) Average Response Time (ART).

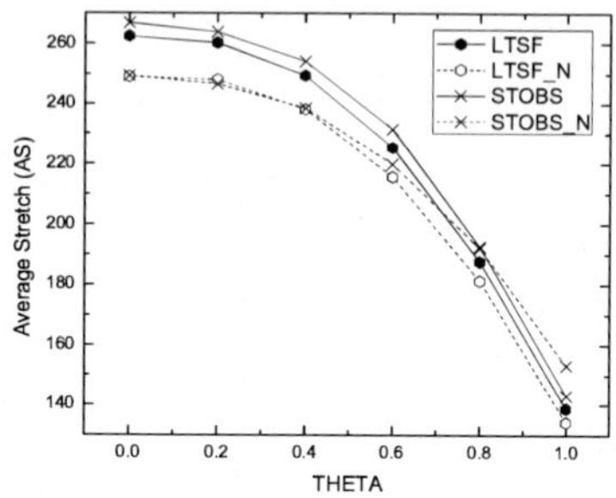

(b) Average Stretch (AS).

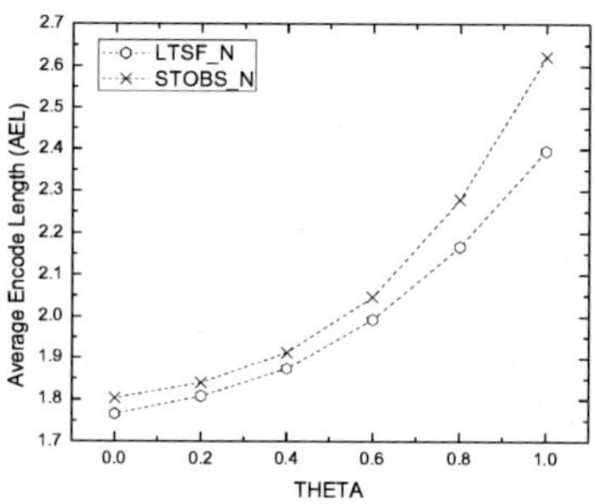

(c) Average Encode Length (AEL).

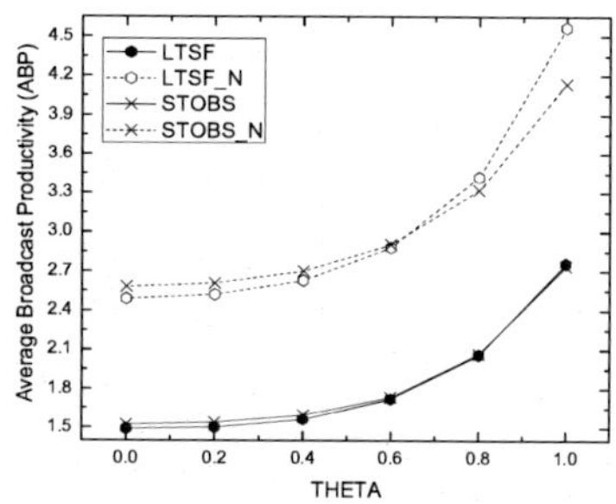

(d) Average Broadcast Productivity (ABP).

Figure 11. Impact of data access pattern (θ) on the stretch optimal model.

Figure 11 exhibits the impact of the data access pattern θ on the performance of the stretch optimal algorithms. When the data access pattern is getting skewed, the AELs of the coding versions increase (Figure 11(c)), and the ABPs of both the non-coding and the coding versions improve (Figure 11(d)). As a result, the ARTs and the ASs of both the coding and non-coding versions reduces (Figures 11(a) and 11(b). However, the amount of improvement in terms of AS (Figures 10(b) and 11(b)) is much less than that of ART (Figures 10(a) and 11(a)) under different setting. In the coding versions of algorithms, data items with different sizes are encoded in a packet and the size of the packet is

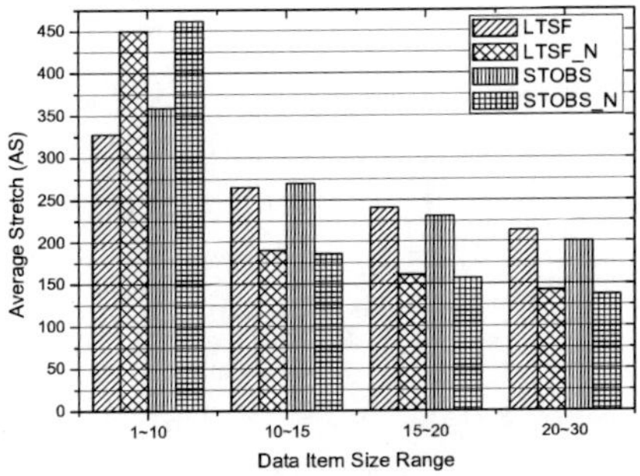

(a) Average stretch (AS) with $\theta = 0.0$.

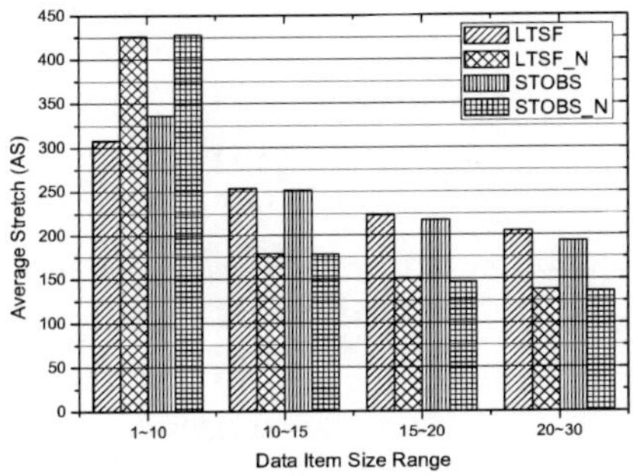

(b) Average stretch (AS) with $\theta = 0.4$.

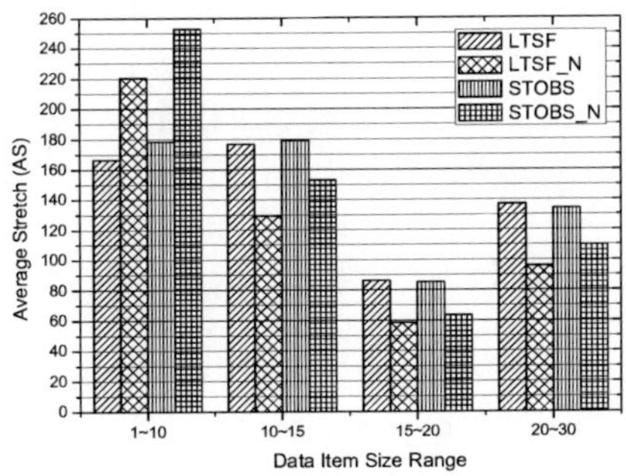

(c) Average Stretch (AS) with $\theta = 1.0$.

Figure 12. Average stretch of different groups of data item size under different data access pattern.

equal to the size of the largest data item. So, the encoded small data items may suffer from long transmission time. Figure 12 shows the ASs of data items with different sizes. Obviously, while larger data items enjoy a smaller AS with network coding, small data items are penalized when using network coding. Since encoded packets contain the data items with different sizes, the performance gain from larger data items is offset by the loss from smaller data items. This explains the slight performance improvement in terms of AS with network coding (Figures 10(b) and 11(b)). In an extreme case, the loss may exceed the gain, leading to worse performance with network coding. For instance, when $\theta = 1.0$ as shown in Figure 11(b), STOBS_N performs worse than STOBS.

6.3.3. Impact of Cache Size

With an increase of cache size, the AEL of the coding version of an algorithm improves (Figure 13(c)). Consequently, the ABP also improves (Figure 13(d)). This helps to reduce both the ART and the AS of a coding version as shown in Figure 13(a) and Figure 13(b), respectively. Figure 13(b) shows that with an increase of cache size, the AS of STOBS_N improves faster that LTSF_N. Figure 14 shows the ASs of different groups of data items under different cache sizes. Specifically, with the small value of *CacheSize* (Figure 14(a)) LTSF_N has a lower AS than STOBS_N for small data items (data items group range 1∼10), and almost equals AS for medium to large data items. In the default setting (*CacheSize* = 60), LTSF_N and STOBS_N have the same AS value for different groups of data items (Figure 14(b)). When the *CacheSize* is getting higher, LTSF_N suffers higher AS than STOBS_N for small data item as shown in Figure 14(c). This implies that LTSF_N penalizes small data items more than STOBS_N by choosing a clique that biases against small data items. Hence, the small data items have to wait longer with LTSF_N, leading to a larger AS.

6.3.4. Impact of Data Item Size

Figure 15 shows the impact of the maximum data item size ($SIZEMAX$) on the performance of the stretch optimal algorithms. The larger value of $SIZEMAX$ means a longer service time (i.e., transmission time) of a data item. Recalling that the response time is the summation of the request waiting time and the item service time. The large $SIZEMAX$ increases the response time in two aspects. First, the transmission time is getting longer. Second, the longer service time of the current data item causes longer waiting time of other data items. As a result,

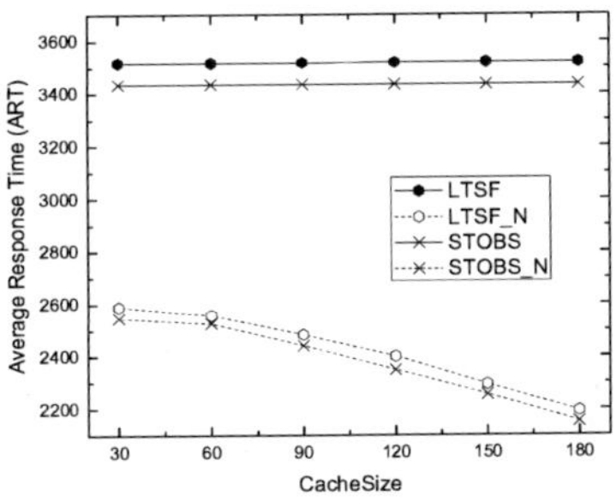

(a) Average Response Time (ART).

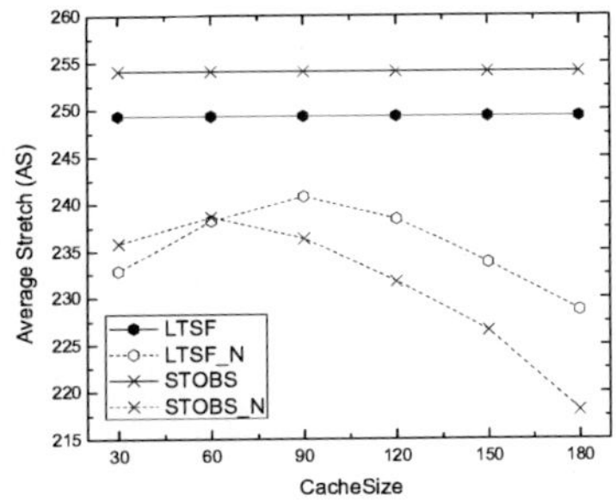

(b) Average Stretch (AS).

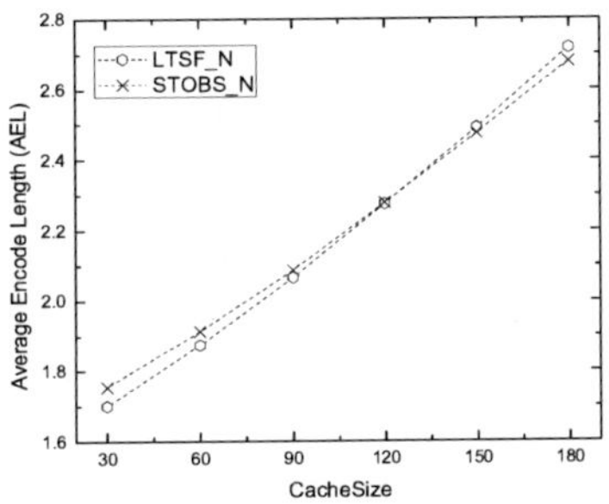

(c) Average Encode Length (AEL).

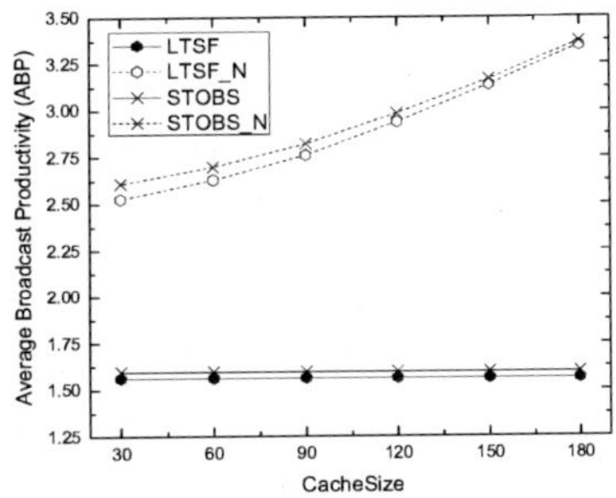

(d) Average Broadcast Productivity (ABP).

Figure 13. Impact of cache size on the stretch optimal model.

with an increase of *SIZEMAX*, the ARTs of both the non-coding and the coding versions increase (Figure 15(a)). Since the stretch is the ratio of the response time to the transmission time of the requested data item, the stretches of the non-coding versions do not changes much because both the response time and the transmission time increase with larger *SIZEMAX* (Figure 15(b)).

In contrast, with an increase value of *SIZEMAX*, the stretches of the coding versions increase. There is a tradeoff of applying network coding in scheduling for different item sizes. On the one hand, an encoded packet may consist of more than one item, which means that broadcasting an encoded packet can satisfy more pending requests simultaneously. Therefore, the average response time decreases. On the other hand, since the transmission time of an encoded

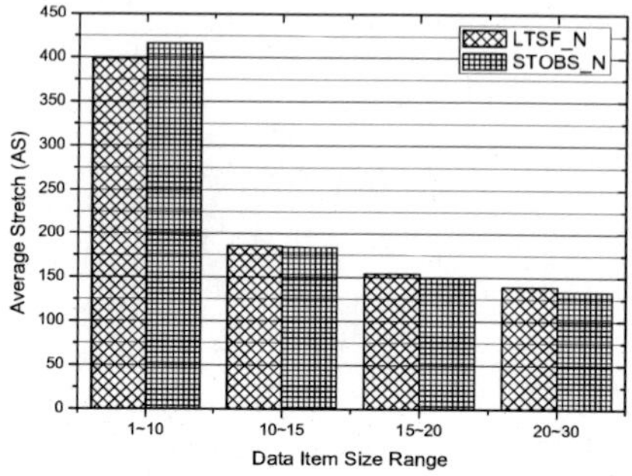

(a) Average stretch (AS) with CacheSize = 30.

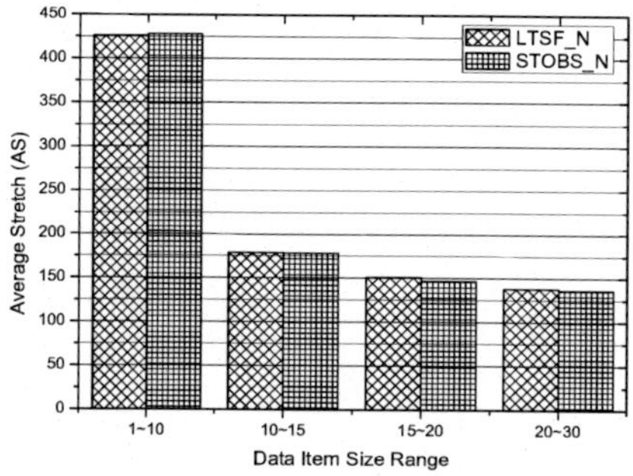

(b) Average stretch (AS) with CacheSize = 60.

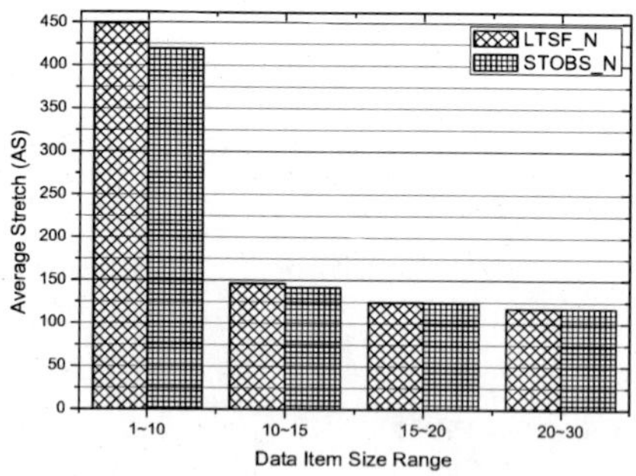

(c) Average Stretch (AS) with CacheSize=180.

Figure 14. Average stretch of different groups of data item size under different cache sizes.

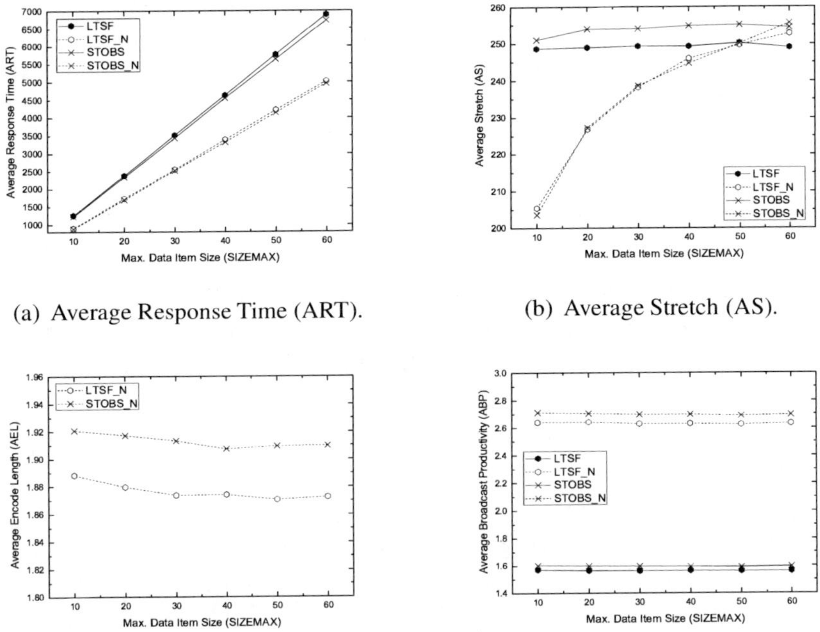

(a) Average Response Time (ART). (b) Average Stretch (AS).

(c) Average Encode Length (AEL). (d) Average Broadcast Productivity (ABP).

Figure 15. Impact of data item size on the stretch optimal model.

packet equals the transmission time of the maximum size item of the corresponding clique, the transmission delay of an encoded packet increases, which results in the longer average response time. To sum up, whether an algorithm can take the benefit of network coding depends on which factors dominates the performance. From Figure 15(a), we note that ART of a coding version is better than its non-coding version under different *SIZEMAX* values. However, in terms of minimizing the stretch, a coding version cannot retain its superiority over the non-coding version when *SIZEMAX* increases to 60 (Figure 15(b)). For instance, when *SIZEMAX* is as low as 10, the coding versions of both LTSF and STOBS improve around 18% stretch over the non-coding versions, however this improvement diminishes very quickly when *SIZEMAX* increases to 40 (3% maximum improvement). Even when *SIZEMAX* reaches to 60, the improvement becomes negative. The reason is due to the penalty of small data items, namely, the coding version of an algorithm incurs higher AS for small data items. The penalty to small data items increases with the value of *SIZEMAX*. Note that *SIZEMAX* have no impact on the AELs (Figure 15(c)). Accordingly, the ABPs of different algorithms are similar under different item sizes (Figure 15(d)).

Conclusion and Future Directions

On-demand data broadcast has been demonstrated as a promising way to disseminate information to a large population of clients in wireless communication environments. Conventional scheduling algorithms cannot best explore the broadcast efficiency of wireless bandwidth. In this study, we apply network coding to enhance the scheduling performance of existing on-demand scheduling algorithms. Based on the derived CR-graph, we propose a heuristic coding based approach, which is able to efficiently transform existing scheduling algorithms to their respective network coding versions while preserving their original scheduling criteria. In particular, based on the classification of on-demand scheduling algorithms, we select a number of representative algorithms in each group for performance evaluation and enhancement, including two real-time algorithms (i.e., EDF and SIN), four non-real-time algorithms (i.e., FCFS, MRF, LWF and RxW) and two stretch optimal algorithms (i.e., LTSF and STOBS). We describe the detailed implementation of the network coding version of each algorithm and investigate their potential performance improvement using a running example. We design a number of metrics and build a simulation model to compare the performance of the non-coding version with the coding version of each algorithm. The comprehensive simulation results under various circumstances conclusively demonstrate that network coding can effectively improve the performance of on-demand scheduling algorithms with different objectives.

In is still our interest to look further into the network coding strategy in heterogeneous service environment where data items are with different sizes. Specifically, since the benefit of network coding largely depends on the size of the encoded packet, which is determined by the largest size of data item which is encoded, it is interesting to strike a balance between encoding large and small data items so as to maximize the overall system performance.

REFERENCES

[1] Ji H., Lee V. C., Chow C.-Y., Liu K., and Wu G., "Coding-based cooperative caching in on-demand data broadcast environments," *Information Sciences*, vol. 385-386, pp. 138 – 156, 2017. [Online]. Available: http://www.sciencedirect.com/science/article/pii/S0020025517300245

[2] Liu K., Ng J. K.-Y., Wang J., Lee V. C., Wu W., and Son S. H., "Network-coding-assisted data dissemination via cooperative vehicle-to-vehicle/-infrastructure communications," *IEEE Transactions on Intelligent Transportation Systems*, vol. 17, no. 6, pp. 1509–1520, 2016.

[3] Ali G. G. M. N., Lee V. C. S., Chan E., Li M., Liu K., Lv J., and Chen J., "Admission control-based multichannel data broadcasting for real-time multi-item queries," *IEEE Transactions on Broadcasting*, vol. 60, no. 4, pp. 589–605, Dec 2014.

[4] Aksoy D. and Franklin M., "R×W: a scheduling approach for large-scale on-demand data broadcast," *IEEE/ACM Transactions on Networking*, vol. 7, no. 6, pp. 846–860, 1999.

[5] Liu K. and Lee V. C., "Performance analysis of data scheduling algorithms for multi-item requests in multi-channel broadcast environments," *Journal of Communications Systems*, vol. 23, no. 4, pp. 529–542, 2010.

[6] Ye F., Roy S., and Wang H., "Efficient data dissemination in vehicular ad hoc networks," *IEEE Journal on Selected Areas in Communications*, vol. 30, no. 4, pp. 769–779, May 2012.

[7] Rehman O., Ould-Khaoua M., and Bourdoucen H., "An adaptive relay nodes selection scheme for multi-hop broadcast in {VANETs}," *Computer Communications*, vol. 87, pp. 76 – 90, 2016.

[8] Liu K., Lee V. C. S., Ng J. K.-Y., Chen J., and Son S. H., "Temporal data dissemination in vehicular cyber–physical systems," *IEEE Transactions on Intelligent Transportation Systems*, vol. 15, no. 6, pp. 2419–2431, 2014.

[9] Ali G. G. M. N., Noor-A-Rahim M., Rahman M. A., Ayalew B., Chong P. H. J., and Guan Y. L., "Cooperative cache transfer-based on-demand network coded broadcast in vehicular networks," *ACM Trans.*

Embed. Comput. Syst., vol. 18, no. 4, Jul. 2019. [Online]. Available: https://doi.org/10.1145/3329865

[10] Noor-A-Rahim M., Ali G. G. M. N., Guan Y. L., Ayalew B., Chong P. H. J., and Pesch D., "Broadcast performance analysis and improvements of the lte-v2v autonomous mode at road intersection," *IEEE Transactions on Vehicular Technology*, vol. 68, no. 10, pp. 9359–9369, Oct 2019.

[11] Xu J., Tang X., and Lee W.-C., "Time-critical on-demand data broadcast: Algorithms, analysis, and performance evaluation," *IEEE Transactions on Parallel and Distributed Systems*, vol. 17, no. 1, pp. 3–14, 2006.

[12] Ali G. G. M. N., Lee V. C. S., Meng Y., Chong P. H. J., and Chen J., "Performance analysis of on-demand scheduling with and without network coding in wireless broadcast," *Future Internet*, vol. 11, no. 12, 2019. [Online]. Available: https://www.mdpi.com/1999-5903/11/12/248

[13] Liu K. and Lee V., "On-demand broadcast for multiple-item requests in a multiple-channel environment," *Information Sciences*, vol. 180, no. 22, pp. 4336–4352, 2010.

[14] Chen J., Lee V. C., Liu K., Ali G., and Chan E., "Efficient processing of requests with network coding in on-demand data broadcast environments," *Information Sciences*, vol. 232, no. 0, pp. 27 – 43, 2013.

[15] Ahlswede R., Cai N., Li S.-Y., and Yeung R. W., "Network information flow," *IEEE Transactions on Information Theory*, vol. 46, no. 4, pp. 1204–1216, 2000.

[16] Ho T., Medard M., Koetter R., Karger D. R., Effros M., Shi J., and Leong B., "A random linear network coding approach to multicast," *IEEE Transactions on Information Theory*, vol. 52, no. 10, pp. 4413–4430, Oct 2006.

[17] Hassanabadi B. and Valaee S., "Reliable periodic safety message broadcasting in vanets using network coding," *IEEE Transactions on Wireless Communications*, vol. 13, no. 3, pp. 1284–1297, March 2014.

[18] Qureshi J., Foh C. H., and Cai J., "Online {XOR} packet coding: Efficient single-hop wireless multicasting with low decoding delay," *Computer Communications*, vol. 39, pp. 65 – 77, 2014, research advances and

standardization activities in {WLANsResearch} advances and standardization activities in {WLANs}.

[19] Yan Y., Zhang B., and Yao Z., "Spacetime efficient network coding for wireless multi-hop networks", *Computer Communications*, vol. 73, Part A, pp. 144-156, 2016.

[20] Zhan C., Lee V. C., Wang J., and Xu Y., "Coding-based data broadcast scheduling in on-demand broadcast," *IEEE Transactions on Wireless Communications*, vol. 10, no. 11, pp. 3774–3783, 2011.

[21] Wang Z. and Hassan M., "Blind xor: Low-overhead loss recovery for vehicular safety communications," *IEEE Transactions on Vehicular Technology*, vol. 61, no. 1, pp. 35–45, Jan 2012.

[22] Ali G. G. M. N., Noor-A-Rahim M., Chong P. H. J., and Guan Y. L., "Analysis and improvement of reliability through coding for safety message broadcasting in urban vehicular networks," *IEEE Transactions on Vehicular Technology*, vol. 67, no. 8, pp. 6774–6787, Aug 2018.

[23] G. N. Ali, Chan E., and W. Li, "Supporting real-time multiple data items query in multi-rsu vehicular ad hoc networks (vanets)," *Journal of Systems and Software*, vol. 86, no. 8, pp. 2127 – 2142, 2013.

[24] Liu K., Ng J. K. Y., Lee V. C. S., Son S. H., and Stojmenovic I., "Cooperative data scheduling in hybrid vehicular ad hoc networks: Vanet as a software defined network," *IEEE/ACM Transactions on Networking*, vol. 24, no. 3, pp. 1759–1773, 2016.

[25] Xuan P., Sen S., Gonzalez O., Fernandez J., and Ramamritham K., "Broadcast on demand: Efficient and timely dissemination of data in mobile environments," in *Proceedings of the 3rd IEEE Real-Time Technology and Applications Symposium (RTAS'97)*, 1997, pp. 38–48.

[26] Wong J. and Ammar M. H., "Analysis of broadcast delivery in a videotex system," *Computers, IEEE Transactions on*, vol. 100, no. 9, pp. 863–866, 1985.

[27] Wong J. W., "Broadcast delivery," *Proceedings of the IEEE*, vol. 76, no. 12, pp. 1566–1577, 1988.

[28] Sharaf M. A. and Chrysanthis P. K., "On-demand broadcast: New challenges and scheduling algorithms," in *Proc. of the 1st Hellenic Data Management Symposium (HDMS'02)*, 2002.

[29] Acharya S. and Muthukrishnan S., "Scheduling on-demand broadcasts: New metrics and algorithms," in *Proceedings of the 4th annual ACM/IEEE international conference on Mobile computing and networking (MobiCom'98)*, 1998, pp. 43–54.

[30] Chaudhry M. A. R. and Sprintson A., "Efficient algorithms for index coding," in *INFOCOM Workshops 2008, IEEE*, April 2008, pp. 1–4.

[31] Karp R., "Reducibility among combinatorial problems," in *Complexity of Computer Computations*, ser. The IBM Research Symposia Series, R. Miller, J. Thatcher, and J. Bohlinger, Eds. Springer US, 1972, pp. 85–103.

[32] Östergård P. R., "A fast algorithm for the maximum clique problem," *Discrete Applied Mathematics*, vol. 120, no. 13, pp. 197-207, 2002, special Issue devoted to the 6th Twente Workshop on Graphs and Combinatorial Optimization.

[33] Hu C.-L., "Fair scheduling for on-demand time-critical data broadcast," in *IEEE International Conference on Communications (ICC'07)*. IEEE, 2007, pp. 5831–5836.

[34] Chen C.-C., Lee C., and Wang S.-C., "On optimal scheduling for time-constrained services in multi-channel data dissemination systems," *Information Systems*, vol. 34, no. 1, pp. 164–177, 2009.

[35] Chen J., Lee V., and Liu K., "On the performance of real-time multi-item request scheduling in data broadcast environments," *Journal of Systems and Software*, vol. 83, no. 8, pp. 1337–1345, 2010.

[36] Ali G. G. M. N., Chan E., and Li W., "Two-step joint scheduling scheme for road side units (RSUs)-based vehicular ad hocnetworks (VANETs)," in *Databse System for Advanced Applications Workshop (DASFAA'11)*, Hong Kong, April 2011, pp. 453–464.

[37] He P., Shen H., and Tian H., "On-demand data broadcast with deadlines for avoiding conflicts in wireless networks," *Journal of Systems and Software*, vol. 103, pp. 118 – 127, 2015.

[38] Polatoglou M., Nicopolitidis P., and Papadimitriou G. I., "On low-complexity adaptive wireless push-based data broadcasting," *International Journal of Communication Systems*, vol. 27, no. 1, pp. 194–200, 2014.

[39] Lu Z., Wu W., and Fu B., "Optimal data retrieval scheduling in the multichannel wireless broadcast environments," *IEEE Transactions on Computers*, vol. 62, no. 12, pp. 2427–2439, Dec 2013.

[40] Wu Y. and Cao G., "Stretch-optimal scheduling for on-demand data broadcasts," in *Proceedings of the 10th International Conference on Computer, Communications and Networks (ICCCN'01)*. IEEE, 2001, pp. 500–504.

[41] Lee V. C., Wu X., and Ng J. K.-Y., "Scheduling real-time requests in on-demand data broadcast environments," *Real-Time Syst.*, vol. 34, no. 2, pp. 83–99, Oct. 2006.

[42] Ploumidis M., Pappas N., Siris V. A., and Traganitis A., "On the performance of network coding and forwarding schemes with different degrees of redundancy for wireless mesh networks," *Computer Communications*, vol. 72, pp. 49 – 62, 2015.

[43] Qu Y., Dong C., Chen C., Wang H., Tian C., and Tang S., "Dcnc: throughput maximization via delay controlled network coding for wireless mesh networks," *Wireless Communications and Mobile Computing*, vol. 16, no. 2, pp. 137–149, 2014.

[44] Chu C.-H., Yang D.-N., and Chen M.-S., "Multi-data delivery based on network coding in on-demand broadcast," in *Proceedings of the The Ninth International Conference on Mobile Data Management*, 2008, pp. 181–188.

[45] Gao Z., Xiang W., Tan G., Yao N., and Li P., "An enhanced xor-based scheme for wireless packet retransmission problem," *Int. J. Commun. Syst.*, vol. 27, no. 12, pp. 3657–3675, Dec. 2014.

[46] Ma Y., Li W., Fan P., Letaief K. B., and Liu X., "On the characteristics of queueing and scheduling at encoding nodes for network coding," *International Journal of Communication Systems*, vol. 22, no. 6, pp. 755–772, 2009.

[47] Chen J., Lee V. C. S., and Zhan C., "Efficient processing of real-time multi-item requests with network coding in on-demand broadcast environments," in *2009 15th IEEE International Conference on Embedded and Real-Time Computing Systems and Applications*, Aug 2009, pp. 119–128.

[48] Sorour S. and Valaee S., "On minimizing broadcast completion delay for instantly decodable network coding," in *IEEE International Conference on Communications (ICC'10)*. IEEE, 2010, pp. 1–5.

[49] Lee V. and Liu K., "Scheduling time-critical requests for multiple data objects in on-demand broadcast," *Concurrency and Computation: Practice and Experience*, vol. 22, no. 15, pp. 2124–2143, 2010.

[50] Schwetman H., "Csim19: a powerful tool for building system models," in *Proceedings of the 33nd Conference on Winter Simulation (WSC'01)*. IEEE, 2001, pp. 250–255.

[51] Zipf G., *Human Behavior and the Principle of Least Effort: An Introduction to Human Ecology*. Cambridge, Mass.: Addison-Wesley Press, 1949.

[52] Xu J., Hu Q., Lee W.-C., and Lee D. L., "Performance evaluation of an optimal cache replacement policy for wireless data dissemination," *IEEE Transactions on Knowledge and Data Engineering*, vol. 16, no. 1, pp. 125–139, 2004.

[53] Ali G. N., Chong P. H. J., Samantha S. K., and Chan E., "Efficient data dissemination in cooperative multi-rsu vehicular ad hoc networks (vanets)," *Journal of Systems and Software*, vol. 117, pp. 508 – 527, 2016.

[54] Ali G. G. M. N., Chan E., and Li W., "On scheduling real-time multi-item queries in multi-rsu vehicular ad hoc networks (vanets)," in *Advanced Information Networking and Applications Workshops (WAINA), 2013 27th International Conference on*, March 2013, pp. 7–12.

In: Vehicular Ad-Hoc Networks ISBN: 978-1-53618-038-1
Editors: G. G. Md. Nawaz Ali et al.

Chapter 4

FAULT DIAGNOSIS TECHNIQUES FOR MOBILE WIRELESS NETWORKS

Hazim Jarrah[1,*]***, G. G. Md. Nawaz Ali***[2]
and Peter H. J. Chong[1]
[1]Auckland University of Technology, Auckland, New Zealand
[2]University of Charleston, WV, US

ABSTRACT

The development of dependable systems has always been of interest and challenge. Generally, the dependability of a system exhibits, to what extent; it can deliver correct services to the users, avoiding frequent and severe service failures. Mobile wireless networks such as MANETs (Mobile Ad Hoc Networks) and VANETS (Vehicular Ad Hoc Networks) provide important services in critical situations. Hence, achieving the dependability in such systems is of most interest and challenging at the same time. This chapter, first, investigates the diagnosis requirements of mobile wireless networks. Also, it describes the main impairments of the dependability in such networks, including faults, errors, and failures. Second, it describes fault diagnosis as a powerful technique to cope with these impairments. Third, it discusses the most common fault diagnosis techniques that have been employed by fault-tolerant systems.

Particularly, it describes these techniques, their strengths and weaknesses, and it reviews their implementations in mobile wireless networks.

Keywords: dependability, fault diagnosis, mobile wireless networks, fault tolerant, MANETs, VANETs

INTRODUCTION

The term "mobile wireless networks," also called dynamic networks or mobile networks, has been used to describe networks experiencing temporal structure [1]. In other words, their topologies vary continually and intrinsically. These networks include mobile ad-hoc networks (MANETs), vehicular ad hoc networks (VANETs), wireless sensor networks (WSNs), and wireless mesh networks (WMNs), just to name a few [2-5]. These networks experience, to various extents, intrinsic characteristics that have significant impacts on operations and services running over them [6-8].

Unreliable Wireless Connections

In wireless networks, network nodes such as laptops, smartphones, and routers communicate through wireless connections that might be impacted by the surrounding environment. Generally, wireless connections suffer from lower bandwidths, high error rates, and frequent disconnections. Hence, the communication latency may increase because of retransmission, timeout delay, and short disconnections to name but a few.

Dynamic and Arbitrary Network Topology

Wireless networks experience temporary and arbitrary topologies for many reasons. Generally, they are deployed randomly with no presumed infrastructure. This may cause an arbitrary topology. In addition, the network structure changes frequently and randomly due to nodes movement. In these networks, nodes move independently, in different directions with varying speeds, and hence the distances among nodes vary over time, and their connectivity varies accordingly. Moreover, the topology may change because of environmental conditions or contentious jamming.

Resource Limitations

The nodes in mobile wireless networks have limited resources and that include battery power, processor, and memory storage. These limitations are implications of the portability needs of mobile nodes.

In addition, mobile wireless networks are mostly deployed into critical and risky environments such as natural disaster areas, flooded regions, and accidents. These deployment environments reflect the importance of these networks and the crucial services they provide. Nevertheless, it reveals the challenges they experience in such hostile surroundings [9].

Both the intrinsic characteristics of mobile wireless networks and their deploying environments may cause a defect in a network component, including its hardware and software, and that is called a fault [10]. These networks are highly prone to faults that impair their performance and hinder their communications [11]. Faults may diminish the ability of the network to perform the necessary tasks and to deliver the required services (impact on network dependability) [8]. In such systems, there are various sources of faults [7, 12, 13]. One of the widespread sources of faults is software defects such as software bugs and crashes in an operating system process. Another source of faults is hardware malfunctions such as poor connections, bad sectors, and CPU fan overheating. Mobile nodes rely on batteries to provide the power they need. Low battery may affect the ability of a node to perform tasks. However, battery capacity is limited, and frequent charging may not be applicable. Hence, low battery level and battery depletion are prevalent causes of faults in such networks. The multi-hops communications between source and destination nodes and the ever-changing topology may prevent message delivering. Wireless channel problems such as network congestions, link failures, and collisions are highly possible reasons for faults. There are also external sources that may cause faults such as environmental interference, electrical interference, Radio Frequency interference, and physical objects. These sources may weaken or even prevent wireless signals. These external reasons are common noise in mobile wireless networks.

The probability of fault occurrence in mobile wireless networks is very high, and a single fault may affect badly a network [14]. Thus, fault diagnosis is crucial for the successful deployment and operation of mobile wireless networks. However, the intrinsic characteristics of mobile wireless networks impose significant challenges on the diagnosis process, and they need to be considered by the diagnosis approach. In the following, we summarize the diagnosis requirements of mobile wireless networks [10, 15].

Minimum Bandwidth Usage

Given the demands on limited bandwidth, the diagnosis protocols must exchange as few messages as possible to keep bandwidth consumption to the minimum possible. In fact, it is preferable to employ ongoing operations and communications in order to identify nodes, experiencing faulty behaviours.

Distributed Diagnosis

The innate dynamics of these networks prevents nodes from gathering global views of the network or sending the diagnosis information on the air. In addition, these networks, by nature, are fully decentralized, and nodes cooperate with each other to perform specific tasks as they suffer from scarce resources, and the diagnosis process is not an exception. A centralized diagnosis causes bottleneck problems, reduce availability, and impair the scalability. Therefore, it is essential to localize and distribute the diagnosis process.

Adaptable

The network environment is unpredictable and arbitrary. Hence, the diagnosis approach must take into consideration the surrounding circumstances and be able to adapt to the changes. In addition, it should deal with the absence of nodes and be robust.

Scalability

The diagnosis approach must deal with an increasing number of nodes in mobile networks. Nowadays, mobile networks consist of hundreds or even thousands of mobile nodes and the diagnosis approach should be able to scale to these network sizes.

Heterogeneous

These networks may include, unlike nodes in terms of resources and capabilities. The diagnosis approach should be able to consider the differences among nodes.

Self-Diagnosis

These networks, by nature, are self-configurable, and the diagnosing process should minimize human interventions. Therefore, it is of importance to automate the fault diagnosis and support the self-diagnosis of networks.

This chapter provides context, relevance, and background to the system-level fault diagnosis problem considered in this chapter. In this sense, it introduces the reader to this field, describing the essential context needed to understand this research problem. To this end, this chapter presents a fundamental set of definitions and concepts relevant to the system-level fault diagnosis. Mainly, it describes the term dependability, its attributes, and its primary impairments, including faults, errors, and failures. It also describes several techniques commonly used to attain the dependability of systems, coping with these impairments. This chapter discusses fault diagnosis as an essential function in developing dependable systems. Several fault diagnosis techniques have been studied, discussing their strengths, weaknesses, and applications. Comparison approach, one of the most practical diagnosis techniques, has gained much attention in the literature to diagnose various kinds of systems. However, this approach has limitations once it comes to dynamic topology systems such as mobile wireless networks. This chapter provides an extensive review of comparison models proposed in the literature. The literature review presented in this chapter helped in identifying the research gaps in this research area and paved the way for developing comparison-based models for mobile wireless networks.

The rest of the chapter is organized as follows. Section 2 provides an overview of the attributes, the threats, and the means of the dependability. Section 3 reviews the fault diagnosis techniques, describing their advantages and disadvantages. Section 4 reviews the comparison-based system-level fault diagnosis in wireless networks. Section 5 concludes the chapter by discussing several research issues open for further investigations.

DEPENDABILITY: ATTRIBUTES, THREATS, AND MEANS

The development of dependable systems has always been of interest and challenge. Generally, the dependability of a system exhibits, to what extent; it can deliver correct services to the users, avoiding frequent and severe service failures [16]. Figure 1 shows the main concepts related to dependability. The dependability criterion evaluates the system behaviour based on essential attributes, namely, availability, reliability, safety, integrity, and maintainability [16]. It is noteworthy that the consideration level of these attributes might vary based on system requirements.

Availability

This attribute measures the readiness for delivering correct services. It is defined as the probability that a system works correctly at time t. This measurement suits the environments where a steady error-free operation is not crucial. It also can be defined based on the mean time to fail, MTTF, which is the average time that a system stays operational before a failure occurs and the mean time to repair, MTTR, which is the average time required to repair a system as the following.

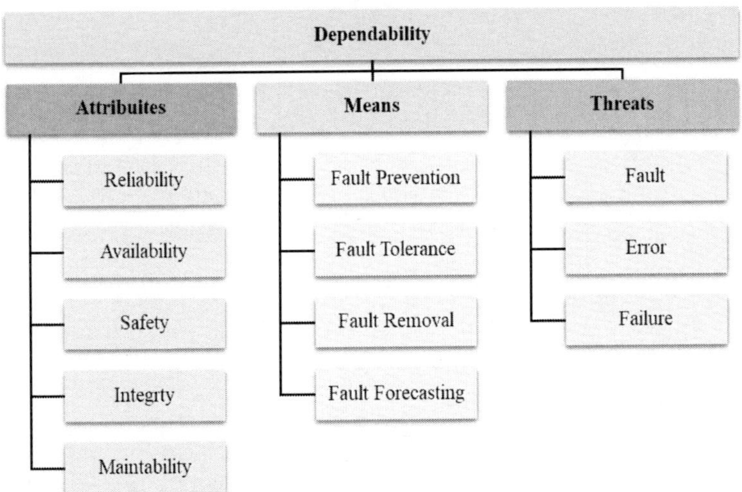

Figure 1. The dependability tree.

$$Availabity = \frac{MTTF}{MTTF+MTTR} \quad (1)$$

Reliability

This attribute measures the continuity of delivering correct services. It is defined as the probability that a system has been continuously operational during the interval $(0, t)$, given that it was operational at time 0. This measurement fits environments, where a singular fault is lethal, and repairing is unavailable.

$$Reliability = 1 - Probability\ of\ Fault \tag{2}$$

Safety

This attribute measures the absence of risks on users and environments. It is defined as the probability of a system, either operates correctly or stops working safely. This measurement is of interest where system failure causes loss of life or environmental disaster.

Integrity

This attribute measures the absence of erroneous system alterations. It is defined as the probability that a system operates safely during its operational time.

Maintainability

This attribute measures the easiness with which a system can be repaired or modified. It is defined as the probability that a failed system can be repaired successfully within a specific time. In other words, the focus here is on decreasing MTTR.

Dependability concept integrates these attributes. However, these attributes themselves are non-orthogonal and tend to conflict with each other sometimes. For example, a high available system might have low reliability if faults occur frequently, and they are handled quickly. Hence, there is always a need to make trades, considering system requirements. In addition, these attributes should be considered in a probabilistic sense since the occurrence of faults is inevitable [16].

Three primary threats may impair the dependability of systems, namely, faults, errors, and failures [17]. A fault is a defect in the design or the operation of a system such as software bug, stuck bits in memory, and omission in data transfer. The fault may lead to an error where the system produces incorrect outputs. The error may provoke a failure where the system delivers incorrect services. For example, a software bug such as an infinite loop in a piece of code is a fault. Once this code is executed, incorrect values (i.e., error) will appear, and hence a crush of the operating system might occur (i.e., failure). It is clear that faults are the sources of other impairments. Figure 2 shows the causality relationship among these events. It is noteworthy that a fault might be dormant, and hence, its impact is unobservable. Once the fault becomes active, it causes an error. In addition, not every error causes failure [6].

There are diverse techniques that have been developed to attain the dependability of systems. These techniques can be classified into four categories [16]: fault prevention, fault tolerance, fault removal, and fault forecasting.

Figure 2. The flow of events.

Fault Prevention Techniques

Fault prevention techniques aim to stop the commencement of faults in a system. These techniques consider avoiding and eliminating the causes of faults into a system before it operates. That is, they employ rigours rules and methodologies during the design processes so that faults are prevented proactively. For example, the network topology can be designed so that the connectivity is maximized. Another example, employing a routing algorithm that considers battery usage to increase the lifetime of the network.

Fault Tolerance Techniques

Fault tolerance techniques aim to deliver correct services sustainably even in the presence of faults. These techniques consider employing the redundancy in a system so that correct services can be delivered without interruption. The redundancy could be in hardware, software, information, and time. Some of the notable examples are fault-masking where nodes cooperate in concealing the impact of a faulty node in a system.

Fault Removal Techniques

Fault removal techniques target reducing the number of faults and their severity. These techniques have been employed to eliminate faults appearing during both system development and operation.

Fault Forecasting Techniques

Fault forecasting techniques anticipate the existing faults in a system and potential faults that may appear in the future and their likely impacts.

These techniques complement each other and not alternatives. In other words, all these techniques must be considered at the design and implementation processes to reach a dependable system that can be trusted [16]. It is noteworthy that the perfection in these techniques themselves is far from realistic, and hence having a fully dependable system might not be achieved. Therefore, and since faults are inevitable, faults tolerance has been of most interest [18]. One of the key steps towards fault tolerance is fault diagnosis that identifies the nature and the cause of faults occurred in a system. Fault diagnosis allows fault-tolerant systems to isolate faulty nodes and prevent them from interrupting service delivery. In the following subsection, we discuss fault diagnosis and its commonly used techniques.

FAULT DIAGNOSIS

Fault diagnosis consists of three main functions, namely fault detection, localization, and identification [19]. Fault detection techniques discover the existence of faults. Fault localization uncovers their locations, and fault identification determines their type and severity. A great deal of research conducted in this area addressed these functions separately. On the other hand, there has been considerable attention to fault diagnosis as one process as well. In this sense, the diagnosis process starts from observing a fault's appearance until having a clear view of the location and the root causes of the fault. It is noteworthy that fault diagnosis and its core functions are the essences of the dependability means [19].

Fault Diagnosis Techniques

This subsection discusses the most common fault diagnosis techniques that have been employed by fault-tolerant systems, as shown in Figure 3. It describes these techniques, their advantages, and disadvantages. In addition, it reviews their implementations in wireless networks.

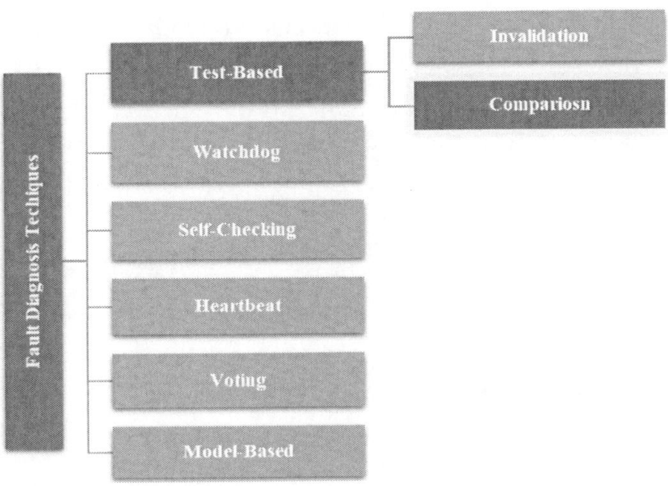

Figure 3. Fault diagnosis techniques.

Test-Based Techniques

These techniques utilize tasks to exercise nodes and identify their conditions. That is, tasks are assigned to nodes in a system, and their test results are used to identify whether they are faulty or fault-free. These techniques can diagnose various types of faults at various levels, and that can be achieved through proper tasks. Test-based techniques can be categorized based on test result interpretations into invalidation-based techniques and comparison-based techniques.

Invalidation-Based Models

In invalidation-based models, nodes test each other directly. That is, node u sends a specific test to node v and compares v's outputs with a set of correct answers. Node v performs the test individually and hence only a fault in v may invalidate the test. Thus, u can identify the fault in v. In the following, we review the main invalidation-based models appeared in the literature.

Preparata et al. [20], introduced the earliest model in 1967 to automate the diagnosis process in multi-processors systems. To this end, they proposed a model, known as the PMC model (symmetric invalidation model), considering the system-level faults in multiprocessor systems. In the PMC model, a system consists of n units where each unit can test other units in the system. The system is represented by a directed graph $G(V,E)$ where the vertex set V represents the units and the edge set E represents the test links. An edge (u,v) is in E if and only if u can test v. Figure 4 shows an example of a testing graph. It is assumed that a unit is either faulty or fault-free, and the unit status does not change during the diagnostic time.

Another diagnostic model was proposed by Barsi et al. [21], called BGM model (Asymmetric invalidation model). The main difference between the PMC model and the BGM model is the behaviour of a faulty tester; the later model reports a mismatching result (i.e., 1) while the earlier model reports an unreliable result. In both cases, the reported results are presented by 0 (pass) or 1 (fail), as shown in Table 1. Both models assume that there is a central observer to diagnose the system as a whole.

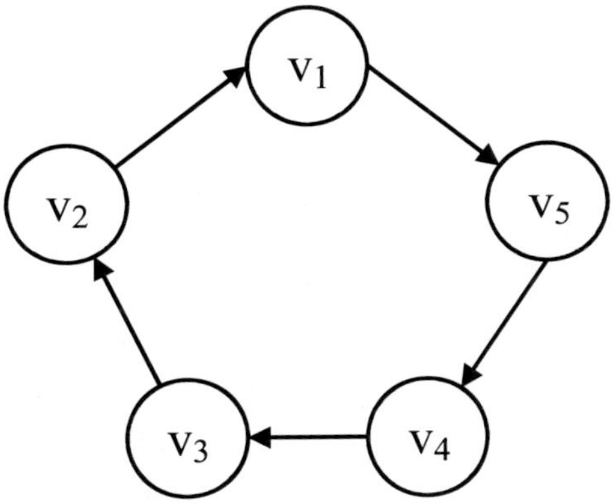

Figure 4. A testing graph example.

Table 1. Symmetric vs. asymmetric invalidation model

Tester	Testee	Symmetric Model	Asymmetric Model
Fault-free	Fault-free	0	0
Fault-free	Faulty	1	1
Faulty	Fault-free	0 or 1	1
Faulty	Faulty	0 or 1	1

Kuhl and Reddy [22] and Kuhl [23] proposed a distributed system-level diagnosis where fault-free nodes cooperate with each other to diagnose the status of all nodes in the system. The earliest distributed diagnosis algorithm that employed this concept is proposed in [22]. Generally, in this model, a node tests its neighbours and maintains a fault vector about their status. This fault vector is, then, broadcasted to neighbour nodes that, in their turn, will test faulty ones in order to confirm their status if they have a direct link.

These models have a predefined set of tests to be executed. In [24, 25], the authors proposed an adaptive diagnosis model where the set of tests are determined based on the results obtained from the previous tests. This model repeats this process until identifying a fault-free node that will act as a tester to identify the status of other nodes. In this sense, this model is a centralized model. Later, a distributed and adaptive model, called Adaptive-DSD, was proposed in [26]. Adaptive-DSD is executed at each node in a system where fault-free nodes test other nodes until finding another fault-free node that will act as a tester and so on. Adaptive distributed diagnosis models suffer from long diagnosis latency and hence, [27, 28] proposed a hierarchical diagnosis to reduce the latency. In hierarchical diagnosis, the diagnosis process is conducted within predefined clusters rather than the whole system.

In [29], the authors proposed using PMC model in WSNs. Particularly, they proposed a testing strategy so that the required diagnosability is met. The strategy uses no reciprocal tests. That is, it does not allow a teste node to test its tester. The suggested strategy divides the network into four quadrants, as shown in Figure 5. Many sensors are involved in the diagnosis process, and hence, this strategy is not energy efficient. In [30], the same authors proposed an improved test assignment approach that takes into consideration energy consumption. Here, the approach also divides the network into four quadrants so that the testing graph includes no reciprocal tests, but they limit the number of sensors involved in the connection assignment. In [31], the authors proposed using a heuristic algorithm to select the optimal number of sensors that take part of the diagnosis process and they considered the distance among nodes for further improvement in terms of energy efficiency.

Invalidation-based models perform direct testing where nodes execute tests between system operations. In fact, designing such tests seems unrealistic because they have to unambiguously, identify whether a node is fault-free or faulty in an acceptable amount of time. In addition, these models consider constructing a testing graph that meets specific assumptions.

Comparison-Based Models

Comparison-based models employ comparison testing instead of direct testing. That is, a tester node assigns the same task to two or more nodes and compares their results to identify faulty nodes, as shown in Figure 6. The idea here is that fault-free nodes behave correctly, and hence, their outputs are consistent.

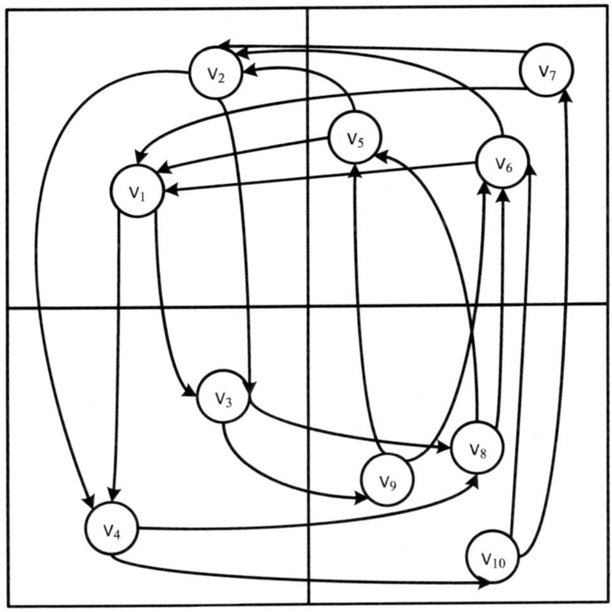

Figure 5. Example of testing region divided in quadrants.

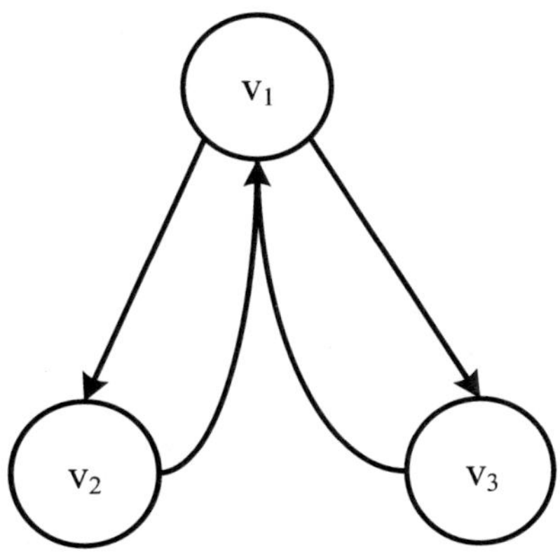

Figure 6. Example of a comparison graph.

Existing comparison-based diagnosis models can be broadly categorized into deterministic or probabilistic models. The former models identify correctly and completely the set of faults in the system, whereas the latter ones offer a correct diagnosis with high probability completeness. The deterministic models, however, impose rigorous requirements on the system's structure and faulty node's behaviour and that limits their usefulness and hinders their scalability.

The earliest comparison-based model is the asymmetric comparison model that was proposed by Malek in 1980 [32]. A year later, Chwa and Hakimi [33] proposed the symmetric comparison model. Figure 7 compares these models when they compare two faulty nodes. Both models assume that there is a central observer to diagnose the whole system. Meang and Malek [34] extended the asymmetric model to create the MM model and assumed that there is a distinct node, which compares the results produced by the nodes. That is, the comparison is distributed and done by the nodes themselves; however, the diagnosis is made by a central node. Sengupta and Dahbura [35] generalised the MM model by allowing the comparator node to be one of the two nodes being compared. The generalised MM (gMM) model same as the MM model sends the comparison results to a central observer.

On the other hand, many distributed models [36] were proposed. In these models, each node tests its neighbours individually, and then, compares their results to produce a local diagnosis view. This view is later disseminated to other nodes to maintain a global view. In these models, the test is assigned from a tester node to a tested node, i.e., one-to-one diagnosis paradigm.

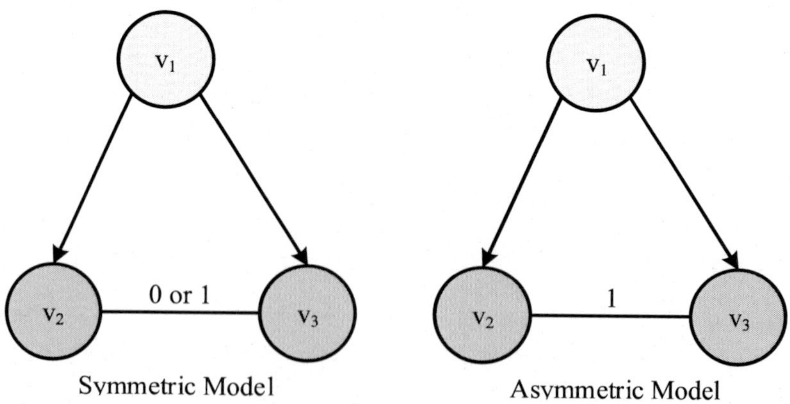

Figure 7. Symmetric model vs asymmetric model where both compared nodes are faulty.

Blough and Brown have proposed a broadcast comparison model (BCM) [37]. This model assumes the existence of a weak reliable broadcast protocol ensuring any message sent by a fault-free node is delivered correctly to all fault-free nodes in the system. When two nodes broadcast their replies, every fault-free node in the system can diagnose them. However, this broadcast protocol seems to be unrealistic for wireless networks. In [38], Chessa and Santi proposed a comparison-based diagnosis model for wireless ad-hoc networks. It exploits the one-to-many communication paradigm to share tests and responses. Hence, bandwidth consumption is reduced. It uses a one-hop reliable broadcast protocol that can be implemented efficiently in ad-hoc networks. While this model is the first model for ad-hoc networks, it assumes a static topology network and static fault model. Elhadef et al. [39] adapted Chessa and Santi's model for time-varying topology. When a node replies, it includes the test task along with the test result, and therefore, any nodes receive the reply message can diagnose the fault. However, the system assumes a static fault model. In [40], Jarrah et al. proposed a time-free comparison model that can diagnose mobile networks with more realistic assumptions. Unlike other models, the time-free comparison model uses no timers and requires no synchronisation.

Dahbura et al. [41] introduced the first probabilistic comparison-based model. This model assumes that a faulty node produces a correct output with a probability, p. Also, it assumes that the number of possible incorrect outputs, m is extremely large. Therefore, the outcome of comparing the output of a faulty node and a fault-free one is match with probability, p. Comparing the outputs of two faulty nodes, however, produces a match with probability, p^2. In [42], Pelc proposed the (p, k)-probabilistic model that utilizes tasks with k possible incorrect outputs to identify the faulty nodes. So, the outcome of comparing the output of a faulty node and a fault-free node or two faulty nodes is match with probability, $q = 1/k$. This model suits tasks with known possible outputs. In [43], Rangarajan and Fussel proposed a probabilistic model that considers the same assumptions of Dahbura et al. model [41]. However, it runs tasks on a small set of nodes, and no global syndrome is analysed. In other words, the diagnosis process is executed for local nodes. These probabilistic models consider both permanent and intermittent faults and impose no limit on the number of faulty nodes. However, they proposed for static multiprocessor systems.

Even though the comparison-based distributed diagnosis is desirable for wireless networks, the most majority of existing comparison-based distributed diagnostic models are not suitable for such networks. They require each node to test its neighbours, one by one through the dedicated wired link. Thus, these models are improper for wireless networks because every test will affect the performance of neighbour nodes, and this will reduce the available bandwidth. Two models might profit from the wireless communications to diagnose the system, namely the Chessa and Santi model and the time-free model. Later, we provide an extensive study of these models and their descendants' protocols.

Watchdog Techniques

Watchdog technique is one of the most popular fault detection techniques. The core of this technique requires auxiliary hardware or software to observe the communications among nodes. If a node fails to perform the expected communications within a specific threshold, then it is considered to be faulty. Due to its simplicity and inexpensiveness, this technique has been used at all levels of system design and for different purposes. In wireless networks, it has been used widely to identify misbehaving and faulty nodes, including selfish nodes that do not cooperate with others, and blackholes, where incoming and outgoing traffics are dropped.

The first paper that introduced the watchdog technique to wireless networks was in [44]. This paper proposed a watchdog mechanism that identifies misbehaving nodes in MANETs. This mechanism has been implemented using two buffers and a timeout timer. The first buffer includes the packets sent recently and the second buffer includes the packets overheard. If these buffers are mismatched for a particular time longer than the timeout, then the forwarding node is considered misbehaving. However, if they are matched, then the watchdog removes the packets and resets the timer. Figure 8 presents the flowchart of this mechanism. The watchdog must know where the packet is forwarding, and hence, the authors suggested to implement it on top of Dynamic Source Routing, DSR routing protocol. This mechanism suffers from incorrect detection if there are ambiguous collisions, receiver collisions, limited transmission power, false misbehaviour, collusion, and partial dropping. Also, the fault coverage is limited to failing to be forwarding packets. The integrity of timers must be considered since the timer, as a part of a node, might be faulty.

To deal with these problems, many variations of watchdog techniques have been proposed. In [45], the authors proposed two protocols to solve the ambiguous collision of the watchdog technique that may occur in wireless networks. In this sense, this paper investigates the watchdog concept as a MAC layer mechanism. Both protocols alter the four-way handshaking in CSMA/CA so that the watchdog suffers no inference monitoring packet's forwarding. These alterations, however, increase the computation and communication overheads. In [46], a watchdog mechanism for wireless networks, called two-way mutual confirmation watchdog protocol has been introduced. This protocol detects misbehaving nodes employing a mutual confirmation on data forwarding between a watchdog node and a receiver node. However, this protocol imposes overhead and delay on the networks. Also, it is supposed that the topology is fixed and there is a routing protocol that provides next-hop nodes information.

The paper in [47] proposed using Bayesian watchdogs that collaboratively detect selfish nodes and black holes in MANETs. Here, the watchdog mechanism has been integrated with the Bayesian filter to evaluate the status of a system probabilistically. Also, neighbour watchdogs work with each other to enhance the detection accuracy and speed. This protocol disseminates the detection information to the whole network, so it has high communication overhead, and it is energy inefficient. Also, using the Bayesian filter imposes a computational overhead.

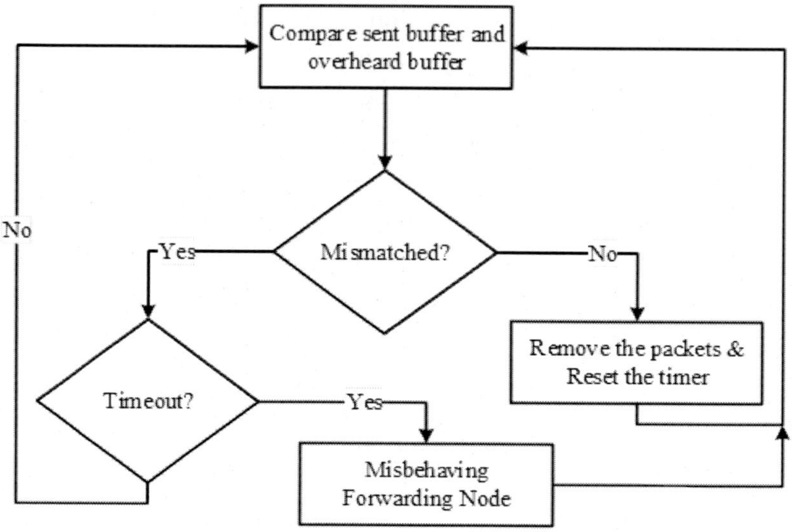

Figure 8. A watchdog-based algorithm flowchart.

In [48], a collaborative contact-based watchdog scheme that detects selfish nodes in delay tolerant networks has been proposed. In this scheme, local watchdogs at each node exchange their detections once the nodes contact. Each watchdog performs two functions: detects selfish nodes and detects new contacts. It also maintains lists of selfish nodes, non-selfish nodes, and the nodes it contacts with them but could not identify their status. This scheme reduces the detection time and increases global detection accuracy. The paper in [49] proposed an enhanced watchdog mechanism that uses an authentication feature among the nodes so that they can authenticate each other for the dissemination of information and opinion among each other.

In [50], a cooperative watchdog method for Byzantine fault detection in wireless networks has been introduced. This method considers cooperation among intermediate nodes and a neighbour watchdog node. The proposed method helps in the existence of Byzantine faults, where erroneous and malicious messages are sent among nodes. The paper in [51] proposed an enhanced watchdog protocol that detects misbehaving nodes while supporting nodes mobility. The proposed protocol employs the Bloom filter, i.e., a probabilistic data structure, that provides authenticating of source nodes. The Bloom filter classifies nodes into fair and malicious nodes using a centralized hash table and validates the nodes through key generation.

To sum up, the watchdog technique has been implemented in various wireless networks, including MANETs, WSNs, and DTNs, to name but a few. The uses of this technique in such networks mainly consider identifying misbehaving nodes based on observing their behaviour [52]. The watchdog-based techniques suffer from false detection due to the holistic behaviour of nodes and the harsh environment they deployed in. Moreover, these techniques consider limited fault models, and they fail to diagnose the root causes of faults.

Self-Checking Techniques

Self-checking techniques help a node to identify its own status by itself. That is, each node in a system tests itself and decides whether it is faulty or fault-free. This has been achieved through various means. Particularly, additional hardware is added to a node architecture to perform self-checking periodically. Alternatively, an algorithm monitors the node behaviour based on specific metrics and identifies its condition. These techniques require no diagnosis messages to be exchanged, and hence, they impose no communication overheads.

These techniques have been used in various wireless networks, exploiting the intelligence of nodes to diagnose themselves. In general, wireless networks are self-organizing and self-healing networks, and as such, nodes adapt to the changes, detecting the ongoing changes in the node itself and the surrounding environments [53].

In [54], the paper proposed using a hardware-based self-detection approach that detects selfish nodes in MANETs. In this approach, a cache hardware unit takes the responsibility to identify misbehaving software and to report that to other nodes in the system.

In [55], the authors designed a circuit using accelerometers that can sense the physical status of a node. This design aimed at detecting physical malfunctions in wireless nodes. In [56], the authors introduced a software-based self-test technique for WSN nodes. Various techniques have been proposed to test the major components of the node, including CPU, memory, and an RF module. The tests have been scheduled so that the test time and energy consumption are reduced.

In [57], the authors proposed a node self-testing approach that uses the data collected by the node itself and its neighbour nodes in WSNs. The self-testing procedures are triggered once certain environmental sensor data is gathered by a node. A fault detected is reported to the base station, which takes the decision of whether to reconfigure or to replace the faulty node.

In [58], the author proposed a software-based self-testing approach for processers embedded in sensors. Self-Tests procedures employ various techniques to reduce energy consumption.

In [59], the author discusses various vehicle self-detection approaches that detect software and hardware faults in a vehicle.

The paper in [60] proposed a self-soft fault detection technique for routing purposes in VANETs. Each vehicle detects its own status, gathering neighbour vehicles information including location, speed, and direction. Nodes experiencing faults exclude themselves from routing paths.

The paper in [61] proposed an aerial-ground cooperative vehicular networking architecture. In which, unmanned aerial vehicles (UAVs) forms an aerial subnetwork, helping the ground vehicular subnetwork through air-to-air (A2A) and air-to-ground (A2G) communications. The UAVs has onboard diagnosis module that provides self-test and energy monitoring. In case of a fault occurring or low battery warning, the UAV requests to return to the ground stations.

In [62], a self-testing mobile communication device has been presented. This device includes a self-testing application stored in non-transistor memory. The self-testing application finds the device location, get a self-test manifest from a server, and run the self-test received.

In [63], the authors presented an invention considering automatic testing of mobile wireless devices. The proposal uses several software agents and environmental conditions to run the most suitable test cases on mobile wireless devices.

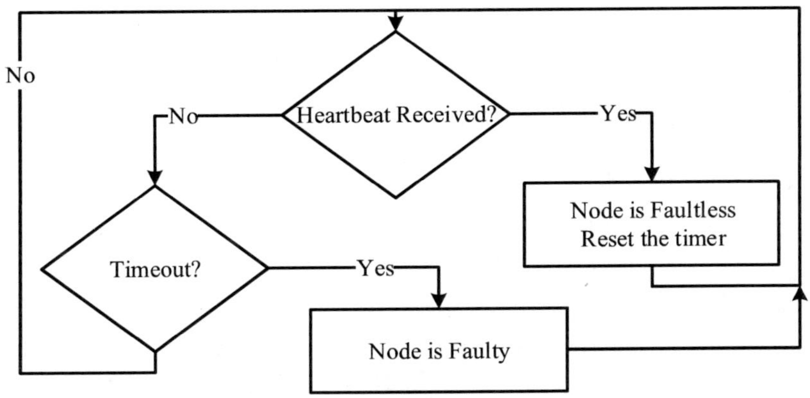

Figure 9. A flowchart of traditional heartbeat technique.

The approach in [64] proposed using a statistical test to identify a node status. That is, each node analyses the data collected from neighbour nodes, employing a z-test statistical. The node then uses its z-score to identify its state. In [65], the same authors proposed using a modified three-sigma edit test to self-detect faults in large scale WSNs.

In [66], each node periodically checks its energy level and informs its manager if the level is below a specific threshold. This requires fewer communications and minimize the response delay of the management system.

Node self-detection techniques implicitly assume that each node is smart enough and it can diagnose itself correctly. This assumption could be unrealistic since faults may prevent the node from performing such a diagnosis since soft-fault may alter the behaviour of the node. In addition, adding auxiliary hardware to node's architecture increase its complexity, weight, and energy consumption and hence, it should be used with caution.

Heartbeat-Based Techniques

Heartbeat-based techniques employ periodic messages, called heartbeats that nodes exchange, informing each other that they are still alive. That is, each node regularly sends an "I am alive" message to its neighbours, confirming that it is still operating. When a node receives no heartbeat messages from one of its neighbours for a specific duration, then it will consider it faulty. These techniques have been used widely to identify nodes, having crash faults. Figure 9 shows the flowchart of a typical heartbeat technique.

The traditional heartbeat technique has two main shortcomings that impact its detection accuracy and latency. First, it expects a fixed timeout delay between two consecutive heartbeats. Second, it relies on the latest heartbeat to detect a crash fault in a node.

These drawbacks have been addressed in various papers [67, 68]. In [69], the authors proposed an improved heartbeat-based technique that uses freshness points where the arrival time of a heartbeat is estimated with respect to the past heartbeats. This technique avoids premature timeouts and improves detection accuracy. In [70], the authors proposed an adaptable timeout duration that takes into account the network quality of service and the application requirements.

It is noteworthy that dealing with the weaknesses of heartbeat technique in wireless and mobile networks is more challenging because of their intrinsic characteristics such as dynamic topology, unreliable links, and asynchronous communications. In [71], a gossip-based failure detection protocol for MANETs is proposed. This protocol uses heartbeats which are represented by counters. That is, each node periodically increases its counter and broadcasts a heartbeat message to its neighbour nodes. The heartbeat messages include a vector of nodes and their latest counter values. Here, also a timer to suspect nodes is set as a fixed value that unfits the dynamic nature of mobile networks. To solve this problem, the paper in [72] proposed adopting the idea of adaptable timeout duration. In [73], it developed an adaptive heartbeat-based failure detector that employs freshness points technique to monitor the status of nodes in wireless ad-hoc and mesh networks. This protocol estimates the arrival time adaptively according to the network load and exchanges heartbeats using a gossip-based protocol. In [74], the authors employed cluster-based communication architecture to detect faulty nodes into each cluster using heartbeats and to forward the detection report across clusters. This solution provides a scalable failure detector for large scale ad hoc networks. In [75], a failure detection protocol for MANET has been proposed. The proposed protocol uses two temporary lists of suspicions and adaptable timeouts. Instead of suspecting a node as faulty, the protocol adds the node to a list of temporary suspicions based on a dynamic timeout, and it starts a second static timeout. If the second timeout expires while no heartbeat has been received, the protocol adds the node to its list of final suspicions. Further improvements on this protocol has been proposed in [76, 77] to improve its detection accuracy and reduce the overhead.

The heartbeat technique has been employed to identify failure nodes in various systems. However, these techniques suffer from several weaknesses. Notably, they diagnose limited types of faults, i.e., crash faults. Also, since nodes depend on periodic messages to diagnose their neighbours, the timing between messages is crucial. Hence, there has always been a trade-off between detection latency and detection accuracy.

Replication with Voting Techniques

Replication has been employed at various level of hardware and software to maintain system fault tolerance. N-Modular redundancy (NMR) is the most popular form of replication. In this form, N similar components perform the same computations and deliver their results to a voter. The voter propagates the most common result to the rest of the system. In other words, the voter masks faults and hide their existence. Figure 10 illustrates NMR. This concept has been applied to various forms of redundancy, such as hardware and software redundancy.

This concept is common in wireless networks where mobile nodes and sensors have been deployed to cover a specific area and to measure or perform specific tasks. In such an environment, the majority of nodes do the same tasks, e.g., run the same software, or collect data. In the field of wireless network, this has also been called neighbour coordination [78].

These techniques rely on the idea that fault-free measurements are spatially correlated, whereas faulty ones are uncorrelated. In these techniques, each node identifies a node state, coordinating with neighbour nodes using majority voting.

In [79], the authors proposed a neighbour coordination-based fault detection algorithm for WSNs. Each sensor node compares its reading with neighbours' readings, considering their confidence level. Also, it counts the number of neighbour sensors which their readings are within a certain threshold. If the majority of readings are within the threshold, the node considers itself to be fault-free. This algorithm has scalability and overhead problems since it goes through several steps, and it needs to exchange results among neighbour nodes several times. It also considers hardware faults only. An improvement for this algorithm has been proposed in [80] where the detection accuracy has been improved even when the number of neighbour nodes is small, and the fault's ratio is high. The proposed algorithm employs an agreement protocol instead of the threshold test.

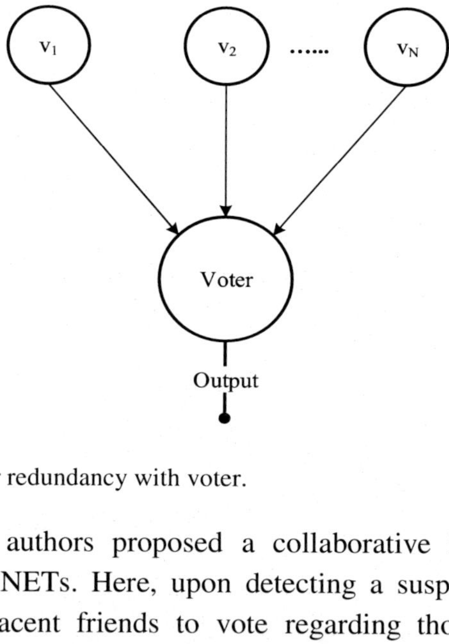

Figure 10. N-modular redundancy with voter.

In [81], The authors proposed a collaborative intrusion detection algorithm for MANETs. Here, upon detecting a suspect node, a source node asks its adjacent friends to vote regarding those events. It then analyzes their votes to confirm whether the suspected node is malicious or not. This algorithm proposed involving friend nodes only in the voting process so that it can trust them.

The paper in [82] proposed two algorithms that identify malicious nodes in MANETS. In these algorithms, a group of nodes votes on the status of suspected nodes and send their votes to a monitor node. The monitor node, then, decides which nodes are malicious based on the votes received. The difference between these two algorithms is that the first algorithm divides the network into cliques, whereas the second algorithm divides it into clusters. A clique is a special cluster where each member in a clique is a neighbour with all other members. The monitor node is assumed to be non-malicious, and it is changed repeatedly. Also, it is assumed that the number of malicious nodes is no more than a specific threshold. To handle these shortcomings, the authors in [83], proposed using trusted voting where only trusted votes are counted. In this algorithm, each cluster includes a header node, monitor nodes, and normal nodes. The header node initiates the detection phase, and the monitor nodes vote to determine the malicious nodes.

In [84], the authors proposed a fault detection scheme for fusion centres in WSNs. The fusion centre is a node that aggregated data from different nodes. The proposed scheme identifies faulty sensor nodes using a majority voting technique. The local decisions of sensor nodes are sent to the fusion centre that considers a node as a faulty node if its local decision is different from the other nodes.

In [85], The authors proposed a weighted-based voting algorithm for WSNs. This algorithm compares the readings of a sensor node with the distance-weighted value of neighbour nodes. In other words, it compares the readings of a node with its neighbour nodes, taking into account the distance among them. If the difference is more than a specific threshold, then the node is faulty. However, the false alarm rate of this algorithm is high in case that many neighbour nodes are faulty.

Generally, the majority voting techniques have been employed in various wireless networks to identify outlier nodes. The fault identification in these techniques depends on the votes of neighbour nodes. Weighted majority voting techniques consider giving weights for votes based on specific parameters. The performance of these techniques is affected by node degree [68].

Model-Based Techniques

The model-based techniques refer to a class of fault diagnosis techniques that consider modelling network components, faults, and events. Some of them consider a network dependency model, storing information about the network components and the relationships among them. A deviation from this model indicates that a fault has occurred. Some of them maintain a fault propagation model that stores information about the possible symptoms and their relationships with faults. The appearance of specific symptoms indicates the occurrence of specific faults. These techniques have been applied recently to wireless networks. The researchers consider many challenges such as obtaining dynamic dependencies models, generalizing the fault and network dependency models, and extracting the root causes given such massive data and models.

In [86], the paper proposed a fault diagnosis architecture and algorithm for MANETs. Their proposal considered the dynamical dependencies in such networks. Also, they proposed a temporal correlation method, associating time with observed symptoms, fault-symptom relationships, and hypothesis of root causes. In addition, they proposed an adaptive fault diagnosis algorithm that identifies the root causes incrementally. An extended version of this paper has been published in [87].

In [88], the authors proposed a system called TimeSAFE that provides a time series analysis, helping the integrated fault correlation system in [89, 90] to identify the reasons for performance degradation.

In [91], the paper proposed using a statistical model to identify whether a wireless network is normal or anomalous. They proposed generating fault signatures for highly occurring faults. Several performance parameters have been collected once a fault happens, and their readings define a fault signature.

In [92], the authors designed a fault detection approach for MANETs. The proposed approach includes collecting various network measurements and statistically aggregating them so that the invariance is ensured. Support vector machine (SVM) has been employed to determine the location and the root causes of faults.

In [93, 94], a fault localization method for service-based systems operated in MANETs, as shown in Figure 11. This method requires architecture to collect and store information needed for the fault localization process. This architecture includes deploying monitors in service components to collect fault symptoms. Also, it incorporates a dependence graph that captures the dependencies among services and a fault propagation pattern that describes the way a fault may lead to failure. Both the fault symptoms and the fault propagation pattern are used to produce the candidate root causes, and finally, a ranking algorithm is used to identify the most likely causes of a fault.

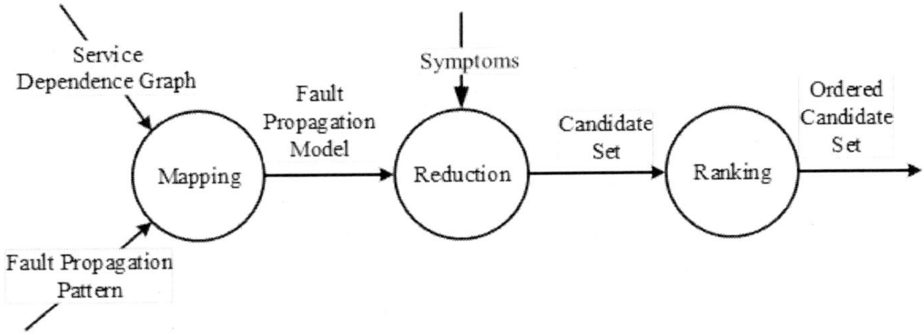

Figure 11. A fault localization method.

There are several drawbacks of model-based techniques. The major drawback of these techniques is the essentiality of having and maintaining accurate, up-to-date dependency models. This is a significant challenge, especially in dynamic environments such as mobile wireless networks. Another drawback is the high computational complexity required to analyze the information and perform the fault localization process. In general, these techniques require collecting information from the system and process them at once.

Concluding Remarks

This subsection compares fault diagnosis techniques based on various perspectives. It also justifies the selection of the comparison approach for further investigations and developments.

Generally, Test-based techniques diagnose faults at various levels, and that can be attained using a test task. In other words, the test task that they employ determines faults of concern. These techniques can identify the root cause of faults, and accordingly, a fault recovery procedure may be initiated. They require no extra hardware to be added to nodes since nodes cooperatively identify the faulty nodes, exchanging diagnosis messages. The diagnosis process in these techniques has been performed both in a distributed and centralized fashion. However, these techniques impose additional communication overhead.

Watchdog-based techniques detect faulty nodes by observing their behaviours. Particularly, nodes observe the communications among neighbours to detect whether a node behaves as expected or abnormally. Generally, they require no extra communications to detect faulty nodes. However, they forfeit the root cause of faults. Moreover, their detection accuracy is affected by the surroundings, and hence, they may provide an incorrect diagnosis in a holistic environment.

Heartbeat-based techniques expect each node to inform other nodes that it is still alive. These techniques have been used to detect crash nodes that fail to update their status for a certain time. They detect faulty nodes relying on exchanging heartbeat messages. As a result, additional communication overhead is inevitable.

Self-checking techniques require each node to detect its own status using either additional hardware integrated into the node, or software including a set of tasks to be performed. These techniques allow each node to diagnose its own faults. However, the additional hardware may increase the node complexity, weight and energy consumption. In addition, a fault may alter the behaviour of a node, impairing its self-checking software.

Majority-voting techniques detect nodes that show oddity comparing with their neighbours. That is, nodes that produce uncorrelated measurements with their neighbour nodes are considered faulty. These techniques have been used where nodes perform the same task. However, these techniques take no account of the root causes of faults. In addition, their performance in sparse networks degrades significantly.

Model-based techniques maintain a network model that describes the components and their relationships. In addition, they maintain a fault propagation model that put aside faults and their potential symptoms. These techniques monitor the events in a network to detect faults. However, building accurate and up-to-date models is a challenge, especially in mobile wireless networks. In addition, analyzing the events to detect faults impose high computational complexity.

It is worth mentioning that there is no individual technique of those described in the previous subsection can satisfy the dependability needs of a system and neither there is a singular technique that suits all systems. As can be seen, every fault diagnosis technique has pros and cons, and there have been extensive efforts to overcome the weaknesses of each technique. Therefore, the focus should be on understating when to use one technique or the other.

Here, we advocate using comparison-based fault diagnosis techniques for mobile wireless networks. The selection of this technique stands on the strengths of the comparison approach, taking into accounts the requirements of mobile wireless networks. Comparison approach is a test-based technique that can diagnose faults and identify their causes. Hence, it is of interest to automate the diagnosis process and initiate a recovery process. Moreover, this approach can identify various types of faults that are of concern.

SYSTEM-LEVEL FAULT DIAGNOSIS

This section investigates the outset of the system-level fault diagnosis problem, its main concepts and the earliest models and techniques proposed to solve it.

System-level diagnosis is a fault tolerance technique, endeavouring to automate the identification of faulty nodes in a system. This technique provides a general concept that could be employed to identify any kind of faults at various levels, and this is implementing through test tasks. Test tasks are systematic drills destined to uncover active faults in a system. For example, writing to the memory and reading the data back to check if there is a fault in the memory. Another example is asking the processor to execute a certain number of arithmetic operations and then take the checksum of the outputs to check if the processor is working correctly. In addition, a node may send a self-message to check if it can receive its own message.

Faults Classification under System-Level Diagnosis

System-level diagnosis considers fault in a system according to several criteria, as shown in Figure 12. On the basis of its persistence, the fault may be classified as permanent, intermittent, or transient. A transient fault disappears eventually without any intervention. If it reoccurs continually, the fault is considered to be intermittent. A permanent fault, however, needs external intervention to remove it. The second criterion is the behaviour of the failed component. If a node cannot communicate with other nodes, the fault is called hard. If a node can communicate while showing an altered behaviour, the fault is called soft. The final criterion is based on the timing of a fault. The fault is said to be dynamic if it is allowed to occur during the diagnosis sessions; otherwise, it is static.

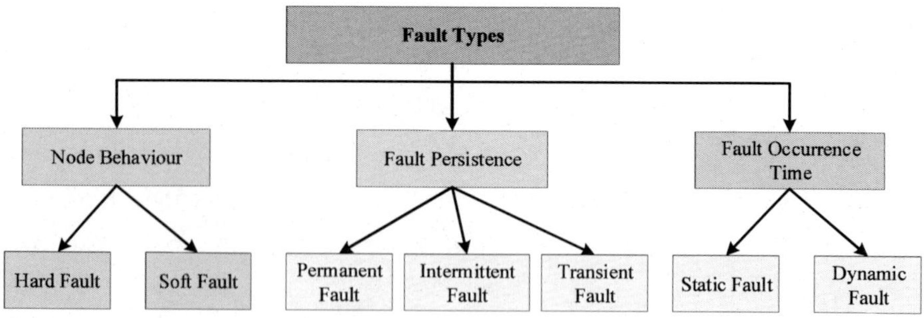

Figure 12. Fault classification.

System-Level Diagnosis Terminologies and Concepts

The research works in the system-level diagnosis problem have considered one or more of the following fundamental problems, namely, the characterization problem, the diagnosability problem, and the diagnosis problem. The characterization problem studies the necessary and sufficient conditions that a system should satisfy in order to allow the specified diagnosis and achieve the required diagnosability. The diagnosability problem investigates the allowable family of fault sets and model that can be diagnosable based on the diagnosis specification. The diagnosis problem identifies the set of faulty nodes in a system based on a given syndrome, fault model, and system model.

The followings terminologies and definitions have been used in connection with system-level diagnosis, and they appear throughout this chapter.

- A *syndrome* is the set of all comparison/test results in a system.
- A *fault set*, *F* is a subset of nodes that are faulty in a system, S.
- A *fault-free set*, *FF* is a subset of nodes that are fault-free in a system, S.
- *Correct diagnosis.* Diagnosis is correct if no fault-free node is identified as faulty. If a fault-free node is identified as faulty, then the diagnosis is an incorrect diagnosis.
- *Complete diagnosis.* Diagnosis is complete if all faulty nodes are identified as faulty. Otherwise, it is an incomplete diagnosis.
- *Deterministic diagnosis*. A diagnosis is said to be deterministic diagnosis if it offers a complete and correct diagnosis. In other words, deterministic diagnosis identifies the set of fault-free and the set of faulty nodes correctly.
- *Probabilistic diagnosis.* A diagnosis is said to be a probabilistic diagnosis if it provides a correct diagnosis with probability completeness. That is, the diagnosis is correct but incomplete.
- *Diagnosis latency.* The time required to complete the diagnosis of a system.

- *Diagnosis complexity*. The number of diagnostic messages exchanged during the diagnosis session.
- *Detection accuracy*. The ratio of faulty nodes detected to the actual number of faulty nodes in a system.
- *False alarm rate*. The ratio of fault-free nodes diagnosed as faulty to the actual number of fault-free nodes in a system.
- *Distributed diagnosis*. Fault-free nodes agree on the set of faulty nodes in a system and isolate them.

Comparison-Based Fault Diagnosis for Wireless Networks

Chessa and Santi introduce a comparison-based diagnostic model for ad-hoc networks [38]. Their seminal work addresses the problem of system-level fault diagnosis in ad-hoc networks. This model utilises the one-to-many communication paradigm to share tests and test responses among adjacent nodes, and hence, reduce the bandwidth required to perform the comparisons [38]. Following their model, many variants have been proposed.

In [40], a Time-free comparison-based diagnostic model has been proposed. This model considers more realistic assumptions that fit mobile wireless networks. Particularly, it tolerates topology changes, it imposes no known bounds on time delays, and it requires no global knowledge about the system under consideration.

To the best of our knowledge, these are the main comparison-based models that have been proposed for wireless networks. In the rest of this section, we describe these models and their protocols.

Chessa and Santi Diagnostic Model

In this model [38], a system comprises n nodes communicating with each other using a one-hop broadcast. At any given time t, the communication can be represented by an undirected graph $G(V, L(t))$, where the vertices, V, are always the nodes, and L is the set of logical links between the nodes at time t. If there is a logical link between two nodes, then they are considered neighbours and are connected by an edge at that time. Each node in the system can be either faulty or fault-free. Faults are permanent and can be either hard or soft.

To diagnose the system, a fault-free node u sends a test request message $m = (u, i, T_i)$ to its neighbours and starts the timer. A timeout should be chosen carefully to give the neighbours enough time to respond. Any node v in N_u reply by a test response message $m' = (u, i, R_v^i)$, Once nodes reply and the timeout expires, the diagnosis process is accomplished by comparing their outcomes using the invalidation rule of the gMM model in Table 2. In particular, no responding nodes are considered to be experiencing a hard-fault. To explain the comparison approach based on this model, assume that a node w in N_v has received m' then two main cases might happen: (1) $w = u$. i.e., the response has been received by the testing node itself. In this case, the node u compares R_v^i with the expected result R_u^i and computes the comparison outcome. If the outcome is 0, then v is fault-free; otherwise, v has a soft-fault. (2) $w \neq u$. and hence (a) $w \in N_u$ or (b) $w \notin N_u$. If $w \in N_u$ then w has received m from u, see Figure 13 (a). Therefore, it can compare R_v^i with R_w^i and conclude the faulty state of v. On the other hand, if $w \notin N_u$, then w should have at least two neighbours in common with u in order to identify the faulty status of these nodes; otherwise, it stores this response, see Figure 13 (b).

Table 2. The invalidation rule of the gMM model [35]

u	v	w	u's Comparison outcome
Fault-free	Fault-free	Fault-free	0
Fault-free	Faulty	Fault-free	1
Fault-free	Fault-free	Faulty	1
Fault-free	Faulty	Faulty	1
Faulty	Any	Any	0 or 1

Many fault diagnosis protocols, which have adopted the comparison approach, are proposed for wireless networks. These protocols are based on Chessa and Santi diagnostic model, which exploits the one-to-many communication paradigm in wireless networks to share tests and test responses. We classified these protocols depending on their dissemination mechanisms into three categories: (a) flooding-based protocols, (b) spanning-tree-based protocols, and (c) clustering-based protocols, as shown in Figure 14.

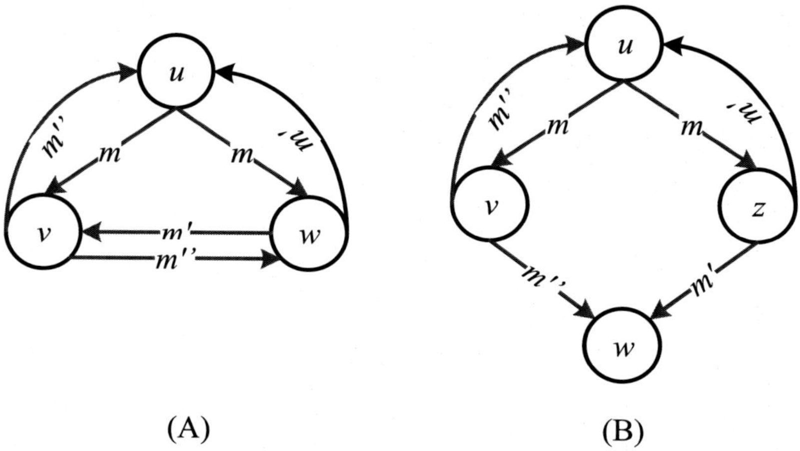

Figure 13. (A) w received u's test task and v's corresponding response reply, (B) w receives u and z's response messages corresponding to u's task [38].

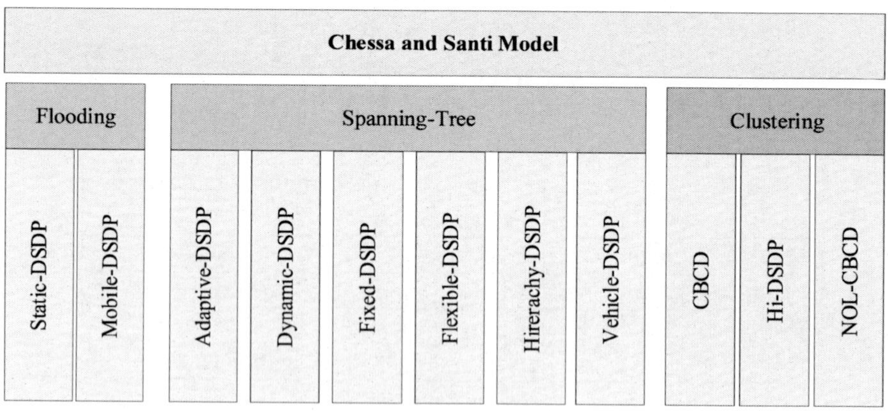

Figure 14. The Chessa and Santi model and its variants.

Flooding-Based Protocols

In these protocols, nodes use a simple flooding mechanism to propagate their local views through networks. With this mechanism, a node sends a message to all of its neighbours and the node, which received the message, resends it to all the other neighbours as long as it is not duplicated. This process goes on until the message is sent to all the nodes in the network. However, this mechanism leads to more redundant messages and thus raises contention and collision problems [95].

Static-DSDP

Chessa and Santi [38] proposed the first comparison-based fault diagnosis protocol for MANETs, called the static-DSDP. This protocol is developed for fixed ad-hoc networks with static, permanent faults. It considers that the nodes are mobile with constraints; hence, the topology has no changes during the diagnosis period. This rather strong assumption is no longer appropriate for MANETs. Thus, Chessa and Santi suggest minimising the diagnosis latency by flooding the local view for every node through the network. Authors provide theoretical correctness proofs and analysis for the protocol with no simulations.

The static-DSDP is composed of two phases, namely, a testing phase and a disseminating phase. The testing phase, for node u, could start either periodically or when any diagnosis message is received from other nodes. At this phase, node u generates and sends a test request message $m = (u, i, T_i)$, where i is the task number and T_i is the task, computes the task's result R_u^i, and starts the timer. Then, node u diagnoses the states of its neighbours to fault-free and soft-fault based on their responses R_v^i, where v is a neighbour node, by using the invalidation rule of the gMM model as in Table 2. On the other hand, the neighbour nodes that do not reply before the timeout time are considered hard-fault. Thus, the timeout time should be selected carefully. After the timeout time, node u enters the disseminating phase and propagates its local view to all the nodes in the network and waits for other nodes' dissemination messages to have a global view about every node in the network. At this time, the static-DSDP protocol terminates for the node u.

This protocol is a timer-based protocol since it relies on timers to identify faults and thus, it makes many timing assumptions on the mobile network. However, these assumptions hinder the protocol's performance in dynamic topology networks. Moreover, selecting a suitable timeout time in mobile networks is a hard problem because nodes' mobility may affect communication delays. A long timeout increases diagnosis latency. On the other hand, a short timeout prevents some neighbour nodes from replying and hence they may be, erroneously, considered faulty.

Mobile-DSDP

In [39], Elhadef, Boukerche and Elkadiki introduce a comparison-based fault diagnosis protocol for time-varying topology MANETs, called the mobile-DSDP. This protocol adapts the Chessa and Santi model. That is, the authors suggest including test task T_i along with test result R_i. This adaption allows a node u that received a response's message m' from node v to diagnose its state even if u was not the tester node. Moreover, in the mobile-DSDP protocol, each node replies to at most $\sigma + 1$ test requests. This suggestion reduces the number of broadcasts, and the nodes will be able to inform at least one fault-free node about their states. Authors provide no simulations for this protocol. However, they prove and analyse its correctness as well as communication and time complexity.

The mobile-DSDP consists of two phases: a testing phase and a disseminating phase. In the testing phase, a node u initiates its diagnosis session by sending a test request message $m = (u, i, T_i)$. Next, it starts two timers T_{out} and $T_{DiagnosisSession}$. Then u collects other nodes test responses until the T_{out} expires and thus, the comparison phase ends. The node u classifies its neighbours into faulty, fault-free or undiagnosed. Moreover, if u received a test response from un-neighbourly nodes, then it can diagnose them either by comparing their results or by doing the task and compares its result with them. At this moment, the node u should maintain three lists: F_u, FF_u, and $UNDIAG_u$. Where F_u represents the list of nodes that were diagnosed as faulty, FF_u is the list of fault-free nodes and $UNDIAG_u$ contains the nodes that have not been diagnosed at the end of the T_{out}. This local diagnostic view, then, is flooded to other nodes into

MANET (disseminating phase). This phase ends in two cases: node u knows the states of all nodes or when the second timer expires. That is, all undiagnosed nodes are classified as hard-fault.

This protocol is, also, a timer-based protocol. Its authors suggest using two timers and an adapted test response message to handle the mobile nodes. The same as the static-DSDP, specifying a suitable timer is really a hard problem. While using an adapted message helps distinguish between hard-fault and migrated nodes, but this may raise resources consumption issues. The more tasks are executed, the more resources are consumed. However, carrying out the task is the last choice. The node rather compares other nodes responses to identify their states, if possible.

Spanning –Tree –Based Protocols

The main idea of this mechanism is to build a spanning tree (ST) in the network, and then nodes propagate and forward diagnosis messages through it. Using a spanning tree leads to a decrease in the number of dissemination messages and rebroadcasts in the network. However, ST, roughly, is very sensitive to topology changes [96].

Dynamic-DSDP

In [97], a comparison-based diagnosis protocol is proposed based on the Chessa and Santi model. This protocol is called the dynamic-DSDP just because it uses a dynamically built ST in the dissemination phase and has nothing to do with a dynamic fault or a dynamic topology MANET.

This protocol consists of two main phases: the testing phase and the dissemination phase. The testing phase may be started periodically by any node or as a reaction to receiving a diagnosis message from another node. In particular, once a node receives a test request, it generates its own test request and starts its timeout time. This phase is similar to that used by the static-DSDP with only a slight difference. The main contribution in this phase is that a node responds to at most $\sigma + 1$ tests, so a node informs at least one fault-free node about its status. In fact, this contribution decreases the number of broadcast response messages. As a consequence of this suggestion, some fault-free nodes may be diagnosed as suspect until the dissemination phase. Once the timeout times expire for all nodes, the dissemination phase begins. This phase has two steps: constructing the ST and disseminating the diagnostic views. The initiator, indicated by the system administrator, builds the ST through propagating ST messages, including IDs of both the node and its parent. Once a node u receives an ST message from v, it checks whether the sender v is a fault-free node, in this case, it sends an ST message to inform its neighbours, including v, that u is the sibling of v. Then, it commences the timer T_{out}. Upon receiving this message, node v adds node u to its children list, and other nodes start their turns. In cases where the timer T_{out} happens before holding a child; then this node is a leaf node. Next, it sends its view to its parent, and each parent collects its children views and sends them to its parent and so on. Then, the initiator aggregates all views into one global view and sends them downward into the ST. notice that the faulty nodes are excluded from the ST.

The main objective of the dynamic-DSDP protocol is to reduce the communication complexity, i.e., the number of exchanged messages. Thus, the authors propose using the ST instead of flooding in the dissemination phase; however, using the ST increases the diagnosis latency. Moreover, they suggest that each node responds to a specific number of test requests. These reductions for the communication complexity make the dynamic-DSDP protocol more energy efficient than the static-DSDP protocol as shown by the authors.

To evaluate the performance of this protocol and compare it with the static-DSDP, the authors implemented both protocols using the NS-2 simulator. They performed a set of experiments to find out how changes in the number of nodes –with all possible faulty nodes- influences the number of messages required to diagnose the system. Also, they used one graph – 80 nodes with 30 nodes connectivity – to study the influence of the number of faults – ranging from 2 to 28 faults – on the communication and time metrics. Moreover, they illustrated how this protocol scales to larger node populations by conducting a set of experiments considering large MANETs with up to 700 nodes.

The simulation results showed that the communication complexity of the dynamic-DSDP is linear and significantly lower than that of the static-DSDP. Moreover, in large-scale MANETs, the results showed the absolute advantage of the dynamic-DSDP over the static-DSDP in term of communication complexity.

Adaptive-DSDP

Elhadef et al. proposed [98] an adaptive-DSDP protocol. This protocol is designed for fixed topology MANETs, the same as the static-DSDP. However, unlike the static-DSDP, this protocol uses an adaptive ST to disseminate nodes' local views rather than the flooding-based dissemination.

The adaptive-DSDP protocol may be divided into four phases: (1) the self-maintaining phase; (2) the testing phase; (3) the self-repairing phase; and (4) the dissemination phase. The adaptive-DSDP assumes the presence of a virtual backbone, a prebuilt ST rooted at an initiator, at the deployment of a MANET. In the maintaining phase, each node, say u, should regularly check its connection with the ST. Once u gets disconnected from its parent, it should reconnect to the ST and find a new parent. At this point, u aims to keep its simple path to the initiator as short as possible; i.e., the lowest depth. This phase terminates as soon as a diagnosis session has been started. The diagnosis session starts either periodically or upon detecting an altered behaviour. Then node u starts its testing phase, which is exactly like the testing phase in the static-DSDP protocol, explained above. After collecting and diagnosing nodes states, the node u initiates the self-repairing phase to make sure that it has a connection to the ST through a fault-free node. If this is not the case, u starts the reconnecting procedure the same as in the self-maintaining phase. Notice that the fault-free nodes exclusively participate and connect to the ST after this phase. Once the ST is repaired, leaf nodes propagate their local views upward to their parents. Each parent collects its children's dissemination messages and sends its view upward as well until reaching the initiator. The initiator accumulates all views in a global view message and then disseminates this message downward through the ST. Thus, every fault-free node knows the states of all nodes in the MANET.

This protocol adopts the ST to disseminate nodes' local views, and thus, reduces the number of broadcasts during the dissemination phase. However, the ST is repaired often, and it could be modified in response to the existence of faulty nodes. This increases diagnosis latency. Figure 15 shows the difference between this protocol and the dynamic-DSDP protocol. Notice that the adaptive-DSDP protocol builds the ST in advance; thus, decreases the protocol overhead. However, both protocols exclude faulty nodes from the ST before disseminating their local views.

The performance of the adaptive-DSDP has been compared to that of both the static-DSDP and the dynamic-DSDP protocol. The authors subjected the adaptive-DSDP to a set of experiments to investigate its time and communication efficiency with regards to the number of nodes and the number of faults. In terms of communication efficiency, the results showed that the adaptive-DSDP exceeds the static-DSDP but falls behind the dynamic-DSDP. That is mainly because the adaptive-DSDP requires, as in static-DSDP, every node to reply for every test during the diagnosis session; but unlike static-DSDP, it uses a pre-constructed spanning tree to disseminate the local views, and thus it has the edge over the static-DSDP.

Hierarchy-DSDP

In [99], Ji and Xu present a comparison-based fault diagnosis protocol for WMNs, called Hierarchy-DSDP. Here, WMN is divided into two levels where level-1 includes mesh clients, and level-2 encompasses mesh routers. This protocol also employs an ST to disseminates the local views of nodes.

Hierarchy-DSDP first diagnoses the mesh routers in level-2 using Chessa and Santi model, assuming they are connected and fixed. Each mesh router, then, ignore faulty routers. Faultless mesh routers are the initiators of the diagnosis process in level-1. That is, a faultless router may initiate the diagnosis session, sending a test request to level-1 clients. Neighbour routers also start their session once they are aware of the commencement of the diagnosis session. When an initiator finishes its diagnosis process, it starts constructing an ST that encompasses the mesh routers. The dissemination phase starts then, level-2 routers exchange their local views through the ST, and each level-2 router generates a global view and sent it to its level-1 clients.

The performance of Hierarchy-DSDP has been evaluated, considering communication overhead and diagnosis latency. It also has been compared with Static-DSDP and Adaptive-DSDP. While it has outperformed the Static-DSDP, it falls behind the Adaptive-DSDP.

Fixed-DSDP

Sahoo and Khilar propose in [100] the fixed-DSDP protocol; a comparison-based fault diagnosis protocol for fixed topology MANETs. It aims to diminish the number of diagnosis messages. Thus, it employs the ST to propagate the local and global views among nodes; furthermore, each node replies for only one test request; moreover, it classifies the nodes into two types: active and passive node; the active nodes, exclusively, create and forward the test task messages.

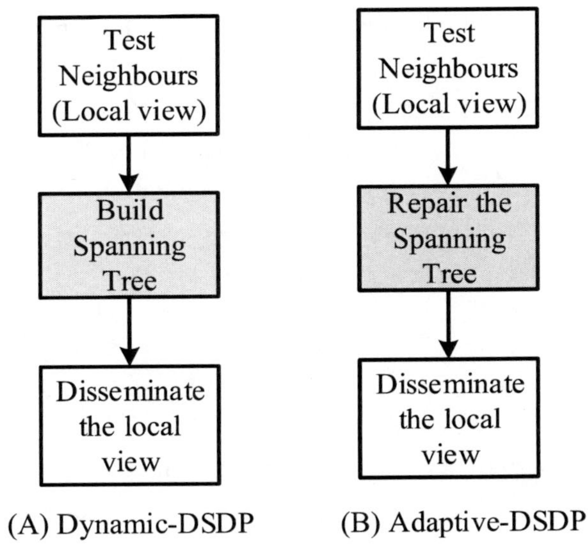

Figure 15. Comparing the dynamic-DSDP protocol and the adaptive-DSDP protocol.

The fixed-DSDP protocol proceeds into three phases called; testing-responding-phase; gathering phase; and building-disseminating phase. Once the diagnosis session initiates, an active node u sends a test-response message to its neighbours. Once received this message, a neighbour node v replies with a message including its test task and its results for both tasks, i.e., its own task and the received task as well. Hereby, it is a test request message and a test response message at the same time. If v is a passive node, then it replies without including its test task. The core of this phase; each node, but the initiator, issue one message. For then, each active node countdowns a timer Tout to handle hard-fault nodes. In the second phase, each active node collects and computes the states of its neighbours and hence maintains its local view. The building-disseminating phase in this protocol is a replica of the one in the dynamic-DSDP protocol. Through this phase, the local views are collected from the leaves to the root and then the global view is generated and disseminated through the opposite direction.

The fixed-DSDP protocol uses a timer to identify the states of nodes and constraints the mobility of the node; thus, hinders the scalability of this protocol. The authors used a java environment (JDK 1.6) to simulate and study the performance of the fixed-DSDP. They investigated the communication efficiency of their protocol in MANETs with a different number of nodes. Also, they compared the proposed protocol with the static-DSDP and the dynamic-DSDP. The simulation results showed that the fixed-DSDP has a linear communication complexity and outperforms both the static-DSDP and the dynamic-DSDP protocol.

Vehicle-DSDP

In [101], Aljeri et al. proposed a fault diagnosis protocol for vehicular networks.

The vehicle-DSDP has four phases as follows. A testing phase is lunched by a gateway that sends a test task to its neighbours. After that, a gathering phase starts aiming to collect the response messages. Once a vehicle receives response messages, it diagnoses the sender as either faulty or faultless according to their results. Any neighbours that did not reply are considered faulty. Next, a building phase constructs an ST of faultless vehicles. That is, an initiator gateway sends a message that triggers the construction of an ST. Here, faulty vehicles are discarded. Lastly, in the dissemination phase, vehicles without children send their local views to their parents. A parent collects its children views and sends its updated view to its parents and so on. Once the initiator collects all of its children views, it generates a global view and sends it back to each road components.

The performance of the vehicle-DSDP has been evaluated using the ns-2 simulator. Two metrics have been measured, namely, the number of packets and the diagnosis latency. Manhattan mobility model has been considered for 300 vehicles and 9 gateways. Two variations of this protocol have been reported. The first considers a single initiator (i.e., the gateway) and the later, also called regional diagnosis protocol, considers each gateway to be an initiator. The results showed that the regional protocol has better performance.

Flexible-DSDP

In [102], Sahoo and Khilar proposed a comparison-based distributed self-diagnosis protocol for MANETs, called Flexible-DSDP. This protocol can detect permanent and intermittent faults. Moreover; it diagnoses dynamic topology networks. Flexible-DSDP maintains and repairs an ST during the diagnosis session to propagate the local views of nodes.

The flexible-DSDP includes four phases, namely, maintenance, comparison, repairing, and dissemination. The maintenance phase considers connecting all the nodes using an ST. In the comparison phase, a node u sends a test task to its neighbours and initiate a timer. Once the timer expires, u generates its local view, classifying the nodes as either faulty or faultless by comparing their results. This phase has to be repeated many times to handle intermittent faults. In the repairing phase, each node assures that it is connected to a fault-free parent based on its local diagnostic view. If that is not the case, it starts a reconnect procedure. Once the ST is repaired, and faulty nodes are excluded, each leaf node sends its local view to its parent. When a parent node receives children local view, it sends its local view to its parent and so on. The initiator then generates and propagates the global view in the ST.

The performance of flexible-DSDP has been evaluated using OMNeT++ simulator. Two performance metrics have been used, namely detection accuracy and false alarm rate. Random waypoint mobility model has been used to study the impact of node mobility on protocol performance.

Clustering-Based Protocols

In cluster-based protocols, nodes are often grouped into clusters. For each cluster, there is a node, which has the responsibility of propagating messages inside and outside the cluster. Using the cluster idea leads to solving a lot of previous protocols' limitations, such as that related to the scalability and node mobility [103]. In general, the employed clustering algorithms affect the efficiency of those protocols.

CBCD

In [104], Li introduces a comparison-based fault diagnosis protocol for a dynamic topology MANETs; called the cluster-based comparison diagnosis (CBCD). This protocol groups nodes into clusters; each cluster has a cluster-head; cluster-heads' neighbours are added to that cluster. The node that appears in more than one cluster is called a gateway.

The CBCD protocol supposes that nodes are grouped, in advance, using the cluster-head-gateway switch routing (CGSR) clustering algorithm in [105]; this protocol also applies the static-DSDP protocol for diagnosing, in front, the states of the cluster-heads.

In CBCD, the diagnosis session activates whenever a cluster-head receives a diagnosis trigger. The cluster-head, then, sends a test request message to its members. Soon after, these members reply by test response's message. Once a gateway node receives a test request, it sends a diagnosis trigger to its neighbouring cluster-heads; thus, gateway nodes play a key role in triggering the diagnosis session at each cluster. After the termination timeout timer, every clusters-head will diagnose each member's states based on Chessa and Santi's model; and maintains a diagnosis information table, which contains two attributes: the host identity HID, and the host state STATE; may have one of four values: fault-free, soft-fault, hard-fault, or uncertain. At the end of this phase, every cluster-head will have its own diagnosis information table about its members. Furthermore, if the diagnosis information table contains nodes whose states are uncertain, it will send a new test request message. Each cluster-head propagates its results to the other cluster-heads via the gateways. The diagnosis session terminates when each cluster-head knows the states of all the nodes in the MANET; any cluster member may ask a cluster-head about the states of the other nodes.

The cluster-heads take the responsibility to generate and propagate both the dissemination message and the test request message. Without a doubt, this decreases the number of broadcasts during the diagnosis session. Another advantage of this protocol is its ability to diagnose nodes that change their location during the first diagnosis session. This is performed by starting a new diagnosis session for the nodes with the uncertain state. However, Using the CGSR algorithm to group the nodes causes some drawbacks such as cluster-head selection, and cluster-head bottleneck. Also, the gateways are diagnosed by each adjacent cluster-head, i.e., more overhead. Moreover, the CBCD protocol depends on the static-DSDP protocol to diagnose the cluster-heads' virtual graph.

The performance of the CBCD protocol has been evaluated and compared with that of the static-DSDP. Two scenarios were simulated, both with 30 nodes. The first scenario considers a fixed topology MANET, while the second scenario supposed that six nodes moved and changed their locations, i.e., dynamic topology MANET. In both scenarios, 12 nodes acted as cluster-heads and gateways, and they did not move. The simulation results showed that the communication efficiency of CBCD is much better than that of the static-DSDP. In addition, the performance of the CBCD protocol under fixed and dynamic topology has been compared, and the simulation results revealed a slightly higher overhead once the topology is dynamic.

Hi-DSDP

In [106], Yadav and Khilar propose a cluster-based fault diagnosis protocol adopting the comparison approach. This protocol uses the hi-DSDP clustering algorithm [27] to group the nodes into clusters; contain $2n$ members, where n is an integer number, see Figure 16.

The hi-DSDP protocol proceeds into two phases: testing phase and dissemination phase; its diagnosis session starts asynchronously from each node; the testing phase proceeds into testing rounds. At the first round, each node tests the cluster that has one node by sending a test request message and countdowns the timer T_{out}. In the second round, each node tests the cluster that has two nodes and starts the timer; and so forth, until testing $N/2$ nodes, Figure 17. Meanwhile, fault-free node and soft-fault nodes reply by test response messages. Once the T_{out} expires, missed nodes are considered hart-fault nodes. Then, each initiator collects its local view; and disseminates it to the other initiator nodes. Finally, each initiator node maintains a global view and broadcasts it.

The hi-DSDP protocol applies a divide-and-conquer testing strategy [27]; thus, nodes perform the diagnosing in a hierarchical way, and the diagnosis information transfers through a tree of depth $\log N$. Hence, diagnosis latency decreases significantly. Nonetheless, it assumes a fully-connected network where every node is connected with all the others. Contrary to this assumption, the sparse network is more realistic and common [107].

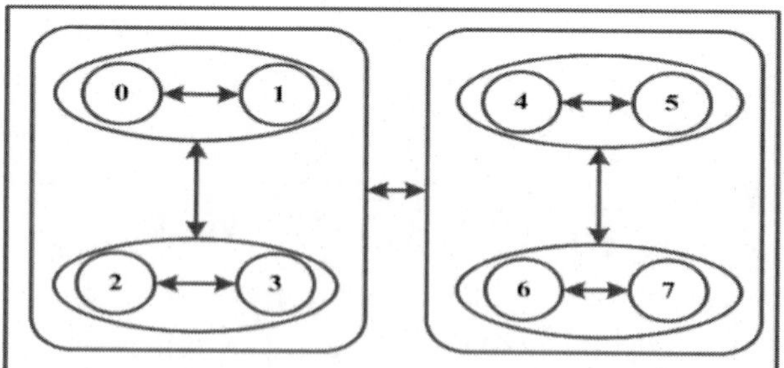

Figure 16. A hierarchical approach to test cluster [27].

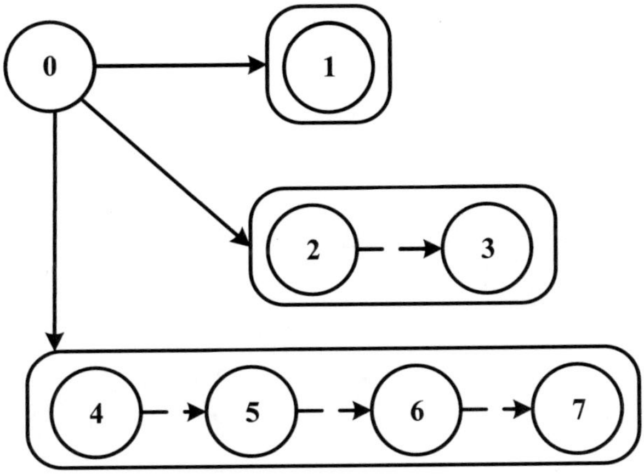

Figure 17. Each node tests all clusters [35].

MATLAB has been used to conduct a set of experiments considering MANETs with varying sizes, 8, 16, 32, 64, and 128 nodes. The authors study how the change in network size affects the time and communication efficiency of their protocol. They also compare this protocol with the CBCD protocol. The results showed that the diagnosis latency for both protocols increases once the network size increases. However, The Hi-DSDP has a significant reduction in the diagnostic latency compare to the CBCD. Concerning the message complexity, the simulation results showed that the Hi-DSDP is linearly scalable and outperforms the CBCD protocol.

NOL-CBDP

In [108], Hassan and Jarrah present a cluster-based fault diagnosis protocol adopting the comparison approach for dynamic topology MANETs. This protocol assumes that a non-overlapping clustering algorithm called 3hBAC [109] maps nodes into disjoint clusters, Figure 18; hence, it employs a spanning tree algorithm to connect the cluster-heads in advance.

In the NOL-CBDP protocol, the cluster-heads have the responsibility of diagnosing the states of their followers. Thus, they should be fault-free to trust their decisions; and hence, the adaptive-DSDP [98] protocol is applied, in front, to diagnose them. Furthermore, the NOL-CBDP proceeds in two phases: a testing phase and a dissemination phase. In the testing phase, cluster-head broadcasts a test request message; starts a timer, and triggers its children in the ST. Thus, each one initiates its diagnosis session. Meanwhile, the members in each cluster reply by a test response message and forward the test request message to their quests, if any. These guests then send back their responses through their access-points to the cluster-heads. Then, each cluster-head diagnosis the states of their members based on the Chessa and Santi model. This local diagnosis view, then, propagates through the ST to reach the initiator cluster-head; which collects its children's views into a global diagnosis view and publishes it downward through the ST.

The NOL-CBDP protocol, like other cluster-based protocols, exploits the cluster-heads to generate and propagate the diagnoses messages. Using the 3hBAC algorithm leads to more stable disjoint clusters without gateways and hence less overhead. However, the node with the highest connectivity is elected as a cluster-head and thus increases the load on this node and may affect its energy and hence the robustness of this protocol.

The performance of this protocol has been studied and compared with the CBCD protocol. That is, the authors used the OMNeT++ simulator to simulate the same two scenarios shown under the CBCD protocol. The NOL-CBCD protocol outperforms the CBCD protocol in fixed and dynamic topology scenario in terms of communication complexity as shown by the simulation results.

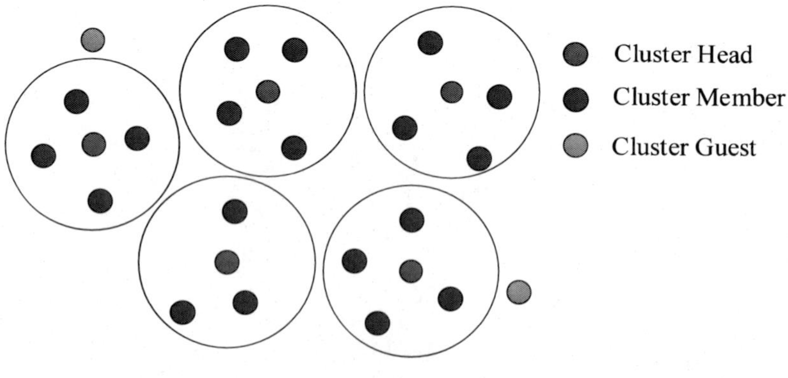

Figure 18. An example of 3hBAC clustering algorithm.

Discussion

Chessa and Santi propose a comparison-based fault diagnostic model that takes advantage of the one-to-many communication paradigm to conduct the comparing process with the least possible communication overhead and time. Many variants protocols are proposed adopting different techniques to disseminate the diagnosis messages.

The flooding-based protocols, namely the static-DSDP and the mobile-DSDP use the simple flooding method to propagate the diagnosis messages. Undoubtedly, flooding produces redundant broadcasts and increases the number of exchanged messages which may affect the energy efficiency of nodes. The main contribution of the mobile-DSDP over the static-DSDP is that mobile-DSDP adapts the response message structure to include the test task along with the test result and hence, it suits dynamic topology networks. However, this method consumes more bandwidth.

The spanning- tree-based protocols use the spanning-tree structure to propagate the diagnosis messages. However, they contrast in two points: when to build the spanning tree and how many tests should the node reply to. However, using the spanning-tree may disrupt these protocols robustness.

The main difference among the clustering-based protocols is the underlying clustering algorithm that has been used to group the nodes. The CBCD protocol uses the CGSR algorithm, the hi-DSDP uses the hi-ADSD algorithm, and the NOL-CBDP uses the 3hBAC clustering algorithm. Generally, these protocols confront the scalability issue. While these protocols outperform the others, the dynamic environment and nodes resources should be taken into consideration when clustering the nodes to make the clustering process more useful and less prone to cause overhead. In fact, the absence of cluster-heads and gateways may affect these protocols' robustness as well.

We summarize in Table 3, Table 4, and Table 5 with a qualitative comparison among the investigated protocols considering the following criteria: (1) Network topology, (2) the dissemination approach, (3) fault duration, (4) fault type, (5) fault time, (6) time complexity, and (7) communication complexity.

Table 3. Network topology and dissemination approach

Protocols	Years	Network Topology		Dissemination Approach
		Static	Dynamic	
Static-DSDP	2001	√		Flooding
Mobile-DSDP	2006	√	√	Flooding
Adaptive-DSDP	2006	√		Spanning Tree
Dynamic-DSDP	2008	√		Spanning Tree
Hierarchy-DSDP	2011	√		Spanning tree
Fixed-DSDP	2013	√		Spanning Tree
Vehicle-DSDP	2013	√	√	Spanning Tree
Flexible-DSDP	2015	√	√	Spanning Tree
CBCD	2007	√	√	Clustering
Hi-DSDP	2010	√	√	Clustering
NOL-CBDP	2011	√	√	Clustering

Table 3 illustrates the investigated protocols with respect to the network topology and the dissemination approach used during the diagnosis session. It is noticeable that all clustering-based protocols consider the dynamic network topology; by clustering the nodes, the stability of network increases significantly [110]. On the other hand, the vast majority of flooding-based and spanning-tree-based protocols are for static topologies. The only notable exception is the mobile-DSDP protocol which proposes sending the test task along with the test result when replying to the test request and hence any node is capable of diagnosing it. While the flexible-DSDP considers dynamic topology networks, its fault detection accuracy under such environments reduces significantly. Also, it implicitly assumes that network topology is static during the comparison phase.

Table 4 shows that all investigated protocols identify the soft or hard, permanent static faults. It is clear that they are ineffective in the presence of dynamic, intermittent, and transient faults. In fact, these kinds of faults appear extensively in dynamic environments like MANET. The investigated protocols make strong assumptions to avoid them and thus limit their contribution to some specific scenarios. The only exception is the flexible-DSDP that considers multiple testing rounds to handle intermittent faults. However, no analysis has been shown for the way to handle these testing rounds and their impact on the performance in terms of time and communication complexity.

Table 4. Supported faults types

Protocols	Fault Duration		Fault Type		Fault Time	
	Permanent	Intermittent	Soft	Hard	Static	Dynamic
Static-DSDP	√		√	√	√	
Mobile-DSDP	√		√	√	√	
Adaptive-DSDP	√		√	√	√	
Dynamic-DSDP	√		√	√	√	
Hierarchy-DSDP	√		√	√	√	
Fixed-DSDP	√		√	√	√	
Vehicle-DSDP	√		√	√	√	
Flexible-DSDP	√	√	√	√	√	√
CBCD	√		√	√	√	
Hi-DSDP	√		√	√	√	
NOL-CBDP	√		√	√	√	

Table 5. The communication and time complexity for each protocol

Protocols	Time Complexity	Communication Complexity
Static-DSDP	$O(\Delta(T_{gen}+T_f)+T_{out})$	$O(n(n+1+d_{max}))$
Mobile-DSDP	$O(\Delta'(T_{gen}+T_f)+T_{out})$	$O(n(n+\sigma+1))$
Adaptive-DSDP	$O(\Delta T_{gen}+(d_{ST}+n-1)T_f+T_{out})$	$O(nd_{max})$
Dynamic-DSDP	$O(\Delta(T_{gen}+d_{ST}T_f)+T_{out})$	$O(n\sigma)$
Hierarchy-DSDP	$O(\Delta T_{gen}+(2d_{ST}+3\Delta'+\Delta_0)T_f+2T_{out}-\Delta_0 T_{gen})$	$O(4n+(n+n_2)d_{max}+2n_2^2-3n_2)$
Fixed-DSDP	$O(\Delta(T_{gen}+d_{ST}T_f)+T_{out})$	$O(n)$
Vehicle-DSDP	*Not provided*	*Not provided*
Flexible-DSDP	*Not provided*	*Not Provided*
CBCD	$O(\Delta(T_{gen}+T_f)+T_{out})$	$O(max(k^2,\zeta k,mk,\zeta m,l,\zeta\eta))$
Hi-DSDP	$O(Nlog^2NC_{i,s}T_{out}+T_{xcg})$	$O(N. C_{i,s})$
NOL-CBDP	$O(d_{ST}T_{gen}+2d_{ST}T_f+T_{out})$	$O(max(kd_{max},k,lq,l,q))$

Table 5 compares the investigated protocols in terms of communication complexity and time complexity. Table 6 provides descriptions for the notations that appear in Table 5. We can see that the cluster-based protocols outperform other protocols and reduce the number of exchanged messages during the diagnosis session. Indeed, this makes them suitable for resource-constrained networks. However, they assume that the nodes are clustered previously. These clusters need reconstruction and maintenance frequently as a consequence of nodes 'mobility and resource limitations

All the investigated protocols are timer-based; they rely on a timer to detect faults; they suppose that message transmission delay and nodes' speed are bounded. However, in dynamic distributed environments like MANETs, these assumptions are hard to implement and limit their usefulness in dynamic topology networks. Additionally, they negatively affect their usage in the presence of dynamic, transient and intermittent faults. Moreover, these protocols assume that all nodes have the same abilities and resources. Definitely, this assumption hinders their ability to be used on heterogeneous networks. Furthermore, the shown protocols consider a connected network with known memberships and structure. Literally, this belief is no longer adequate for dynamic networks.

With regards to the performance evaluations of surveyed protocols, our findings show that theoretical correctness proofs and analysis have been presented for the majority of protocols. Moreover, various simulation tools were used to evaluate the performance of the most majority of the proposed protocols, see Table 7. That is, a combination of simulation and analytical methods was used for performance analysis and evaluation. Nonetheless, the static-DSDP and the mobile-DSDP have no experimental evaluations. However, our survey results showed a lack of well-documented simulation results. That is, relevant information about the simulation tools and parameters has not provided. For example, the name of the simulation tool used for CBCD protocol has been concealed. In addition, the simulation settings and parameters have not been clearly disclosed in many surveyed papers. This raises doubts about the suitability and the reproducibility of these simulation experiments.

Table 6. Notations

Notations	Description
n	The number of nodes
d_{max}	The maximum of the node degrees
Δ	The diameter of G
T_{gen}	The elapsed time between the reception of the first diagnostic message and the generation of the test request
T_f	The time needed to propagate a dissemination message
T_{out}	Timer
Δ'	The maximum of the diameters of graphs G_t
σ	The total number of faulty mobiles
d_{ST}	The depth of the spanning tree
Δ_0	The diameter of the subnet of level-1
n_2	the number of mesh routers in level-2
k	The number of cluster-heads
ζ	The number of diagnosis periods triggered due to the movements of the active mobile hosts.
m	The total number of gateways
l	The total number of cluster members
η	The number of active mobile hosts
q	The number of cluster-quests
N	The total number of initiators
T_{xcg}	The time needed to exchange the diagnosis information by all initiators in the network
$C_{i,s}$	A list of ordered nodes tested by node *i* in a cluster of size 2^{s-1}, in a given testing round.

Whereas the scalability of the diagnosis protocol is one of the main diagnostic requirements for mobile wireless networks, most works demonstrated simulation results for a limited number of nodes. The time efficiency for most surveyed protocols has not been presented. While some might argue that the communication efficiency is more crucial in such networks, we believe that the evaluation of diagnosis latency is required for in-depth understanding and characterising of protocols.

Considering the diagnosis requirements for mobile wireless networks, the surveyed protocols have different design goals. Thus, based on the network under consideration, each protocol has its strengths and weaknesses. In this sense, the flooding-based protocols exchange a large number of messages during the diagnosis session. Thus, they consume much energy since energy consumption is directly proportional to the number of exchanged messages. Moreover, they cannot scale well since the number of exchanged messages is directly related to the number of nodes in the network. That is, flooding based protocols are neither energy-efficient nor scalable. Accordingly, they are unsuitable for large-scale, high density or energy-constraint networks. However, the design goal of these protocols was to minimise the diagnosis latency and terminate the diagnosis session as soon as possible for the sake of a realistic fixed topology assumption. In this sense, these protocols are time-efficient and suitable for dynamic networks. In particular, the mobile-DSDP protocol is recommended for highly dynamic networks because it considers the dynamics of topology during the comparison phase. This protocol could be extended to be more communication efficient by using a minimum spanning tree during the dissemination phase.

Table 7. Experimental evaluations

Protocols	Simulation	Compared with	# of nodes
Static-DSDP	None	None	None
Mobile-DSDP	None	None	None
Adaptive-DSDP	NS-2	Static-DSDP & Dynamic-DSDP	10 -700
Dynamic-DSDP	NS-2	Static-DSDP & Adaptive-DSDP	10-700
Hierarchy-DSDP	Not provided	Static-DSDP & Adaptive-DSDP	10-100
Fixed-DSDP	Java	Static & dynamic-DSDP	10-100
Vehicle-DSDP	NS-2	None	309
Flexible-DSDP	OMNeT++	None	100-1000
CBCD	Not provided	Static-DSDP	30
Hi-DSDP	MATLAB	CBCD	8-128
NOL-CBDP	OMNeT++	CBCD	30

On the contrary, communication efficiency was given precedence over time efficiency in the spanning tree-based protocols. That is, they aimed to reduce the number of exchanged messages and hence reduce energy consumption. Therefore, these protocols are more suitable for energy-limited networks. However, they have rigid assumptions about topology changes and these assumptions hinder protocols usefulness in dynamic networks. Using minimum spanning tree rather than spanning tree could be more appropriate for these protocols. In addition, one might use tree-based algorithms explicitly designed for mobile networks during the dissemination phase as in [111].

In general, surveyed cluster-based protocols offer many advantages for large-scale and mobile networks. In these protocols, specific nodes, cluster heads, are responsible for diagnosing the system. This leads to optimise energy consumption, tolerate node movements, and reduce the number of exchanged messages and diagnosis latency. These protocols could be applied to WMNs [112].

Time-Free Diagnostic Model

In [40], the authors developed a comparison-based fault diagnosis model for mobile networks, called time-free diagnostic model. This model considers a class of diagnosable dynamic systems that experience temporary structures and asynchronous communications. These systems are represented by a time-varying graph, which has a dynamic topology, and the connections among nodes fluctuate randomly.

The time-free diagnostic model utilizes the comparison approach to diagnose the nodes in a system. In this model, nodes cooperate with each other to diagnose the whole system in a distributed fashion. Unlike existing comparison-based model, this model uses no timers; rather, it uses message exchange patterns. This feature is of interest since it implies that the time-free diagnostic model relaxes the traditional constraints. As a result, this model tolerates topology changes and asynchronous communications and hence, it is more suitable for mobile networks.

In this model, a faultless node u sends a test message, including a task to its neighbour nodes. Any node, v, which receives the test message, executes the task and sends a test response message, including the task and the result. Also, v sends its own test message to diagnose its neighbours. When u receives a sufficient number of responses, it compares their results, using the invalidation rule, as shown in Table 2. Then, u generates its partial view that includes the faulty and faultless nodes along with timestamps. The nodes that did not reply are diagnosed as faulty. Partial views may mistakenly diagnose mobile and slow faultless nodes as faulty. Nonetheless, Nodes are supposed to cooperate with each other to update their views and acquire the latest diagnosis. The most crucial parameter in this model is the number of responses, α_u that each node u should wait for. However, α_u can be calculated locally by each node based on the number of neighbours, $|N_u|$ and the maximum number of local faults, f_u as follows. $\alpha_u = |N_u| - f_u$; $|N_u| > 2. f_u$ and $\alpha_u \geq f_u + 1$.

Based on this model, two fault diagnosis protocols have been proposed for mobile networks. These protocols use two different dissemination techniques to exchange partial views among the nodes in a system.

Time-Free-DSDP

In [40], Jarrah et al., propose a time-free comparison-based fault diagnosis protocol for mobile networks. This protocol implements the time-free diagnostic model and uses no timers. Both hard faults, as well as static and dynamic soft faults, can be diagnosed correctly. This protocol has no constraints on topology changes and communication delays.

The time-free-DSDP encompasses two phases: testing phase and dissemination phase. The execution of this protocol starts either periodically or because of an abnormal event. In the testing phase, nodes comply with the time-free diagnostic model so that each faultless node generates a partial diagnosis view by the end of this phase. This partial view includes the list of faulty and faultless nodes from a node perspective. In the dissemination phase, nodes exchange their partial views using an adapted flooding mechanism. That is, faultless nodes propagate the partial view they receive if it has updated information. Later, each node constructs a complete and correct diagnosis view, considering the most recent decisions.

The correctness of the time-free-DSDP has been proved. Moreover, its performance has been analysed with regards to communication complexity and time complexity.

RLNC-DSDP

In [113], Jarrah et al., propose another time-free comparison-based diagnosis protocol for mobile networks, called RLNC-DSDP. This protocol employs a random linear network coding (RLNC) technique [114] during the dissemination phase instead of flooding mechanism to propagate the partial view of nodes.

The RLNC-DSDP has testing phase and dissemination phase. In the testing phase, nodes employ the time-free diagnostic model to diagnose a sufficient number of nodes, generating a partial diagnosis view. Later, these views are exchanged, employing an RLNC technique. In this technique, nodes instead of forwarding the partial views individually, they combine them and send out an encoded packet that encompasses them. Each node collects encoded packets into a decoding matrix. Once this matrix becomes full, a node decodes the packets and retrieve the original ones (partial views). Then, each node aggregates the partial views to create a complete and correct diagnosis view about the system.

The RLNC-DSDP utilizes the network-coding concept to reduce the communication overhead and hence, it has the edge over the time-free-DSDP protocol in mobile networks. The performance of the RLNC-DSDP has been evaluated against the communication overhead and diagnosis time using OMNeT++ simulation. The results show that the RLNC-DSDP is efficient, and the communication overhead is significantly reduced.

Discussion

The time-free diagnostic model takes advantage of the message exchange pattern to implement the comparison approach, relaxing the rigid constraints considered in Chessa and Santi model. Hence, it suits mobile wireless networks that experience topology changes and asynchronous communications. In addition, this model diagnoses dynamic faults as well as static faults. Based on this model, two protocols have been proposed. First, the time-free-DSDP that uses a simple flooding mechanism to exchange the partial view of nodes. Second, RLNC-DSDP that employs RLNC technique to exchange the partial view of nodes. This model and its variants are viable for mobile wireless networks since they respect the requirements of mobile networks.

CONCLUSION

In this chapter, we study the characteristics of mobile wireless networks and their design requirements. These networks provide crucial services, and their successful operations are of most importance. However, they are highly subjected to faults because of their deploying environments and their deep-seated characteristics. Faults diminish network abilities and hinder network dependability. Therefore, they need to be handled. We study the dependability as an indication of the trust we can put on a network. Also, we study the major means employed to enhance the dependability along with the threats and the attributes of the dependability. Since fault is inevitable, fault tolerance techniques are of most interest. Fault diagnosis is the most important step towards dependable systems. Hence, we focus on fault diagnosis techniques that have been implemented in wireless networks. We study these techniques and their pros and cons. Comparison-based fault diagnosis techniques have been advocated to be used in mobile wireless networks in order to diagnose the state of nodes.

Particularly, we study the Chessa and Santi comparison-based model proposed for wireless networks. This model is of interest to our research since it is the seminal model that considers the comparison-based diagnosis process in wireless networks. We surveyed various comparison-based fault diagnosis protocols for wireless networks. In addition, we proposed a taxonomy for categorising these protocols into flooding-based, spanning-tree-based, and clustering-based protocols. We found that the clustering-based protocols perform better than non-clustering protocols. However, cluster formation and maintenance require extra overhead, which may have an impact on system performance. It is suggested that the clusters can be constructed based on some management criteria to improve lifetime.

The traditional models such as Chessa and Santi's and its descendants are timer-based and, therefore, they impose constraints on the system under consideration. Research is required to develop robust diagnostic models to assist fault detection and diagnosis, given the planning and design requirements of such systems. The discussions revealed that the Chessa and Santi model imposes stringent assumptions such as synchronous communications, global knowledge about the networks. These assumptions hinder its usefulness in dynamic topology networks. For example, their protocol fails to provide a correct and complete diagnosis under dynamic topology hypothesis. Elhadef et al. [39] suggest an extension for this model so that the dynamic topology can be tolerated. However, their protocol requires two timers, and hence, its benefits for dynamic topology networks are hard to achieve.

Recently, the time-free diagnostic model has been developed. This model respects the intrinsic characteristics of mobile wireless networks. The potential of this model needs further investigations.

REFERENCES

[1] Casteigts, A; Flocchini, P; Quattrociocchi, W; Santoro, N. "Time-varying graphs and dynamic networks," presented at the *Proceedings of the 10th international conference on Ad-hoc, mobile, and wireless networks*, Paderborn, Germany, 2011.

[2] Yick, J; Mukherjee, B; Ghosal, D. "Wireless sensor network survey," *Computer networks*, vol. 52, pp. 2292-2330, 2008.

[3] Qiu, T; Chen, N; Li, K; Qiao, D; Fu, Z. "Heterogeneous ad hoc networks: Architectures, advances and challenges," *Ad Hoc Networks*, vol. 55, pp. 143-152, 2017.

[4] Rasheed, A; Gillani, S; Ajmal, S; Qayyum, A. "*Vehicular Ad Hoc Network (VANET): A Survey, Challenges, and Applications*," Singapore, 2017, pp. 39-51.

[5] Akyildiz, IF; Wang, X; Wang, W. "Wireless mesh networks: a survey," *Computer networks*, vol. 47, pp. 445-487, 2005.

[6] Zhang, Z; Mehmood, A; Shu, L; Huo, Z; Zhang, Y; Mukherjee, M. "A survey on fault diagnosis in wireless sensor networks," *IEEE Access*, vol. 6, pp. 11349-11364, 2018.

[7] Azar, HN; Malazi, HT. "Decentralized detection of hybrid faults in mobile sensor nodes," *Simulation Modelling Practice and Theory*, vol. 87, pp. 210-225, 2018.

[8] Basile, C; Killijian, MO; Powell, D. "A survey of dependability issues in mobile wireless networks," *Citeseer*, 2003.

[9] Reina, D; Askalani, M; Toral, S; Barrero, F; Asimakopoulou, E; Bessis, N. "A survey on multihop ad hoc networks for disaster response scenarios," *International Journal of Distributed Sensor Networks*, vol. 11, p. 647037, 2015.

[10] Chadha, R; Kant, L. "Fault Management," in *Policy-Driven Mobile Ad hoc Network Management*, ed: John Wiley & Sons, Inc., 2007, pp. 225-273.

[11] Clegg, RG; Clayman, S; Pavlou, G; Mamatas, L; Galis, A. "On the selection of management/monitoring nodes in highly dynamic networks," *IEEE Transactions on Computers*, vol. 62, pp. 1207-1220, 2013.

[12] Raposo, D; Rodrigues, A; Silva, JS; Boavida, F. "A Taxonomy of Faults for Wireless Sensor Networks," *Journal of Network and Systems Management*, vol. 25, pp. 591-611, July 01 2017.

[13] łgorzata Steinder, M; Sethi, AS. "A survey of fault localization techniques in computer networks," *Science of computer programming*, vol. 53, pp. 165-194, 2004.

[14] Mahapatro, A; Khilar, PM. "Fault Diagnosis in Wireless Sensor Networks: A Survey," *Communications Surveys & Tutorials, IEEE*, vol. 15, pp. 2000-2026, 2013.

[15] Jarrah, H; Sarkar, NI; Gutierrez, J. "Comparison-based system-level fault diagnosis protocols for mobile ad-hoc networks: A survey," *Journal of Network and Computer Applications*, vol. 60, pp. 68-81, 2016.

[16] Avizienis, A; Laprie, JC; Randell, B; Landwehr, C. "Basic concepts and taxonomy of dependable and secure computing," *IEEE*

transactions on dependable and secure computing, vol. 1, pp. 11-33, 2004.

[17] Avižienis, A; Laprie, JC; Randell, B. "Dependability and its threats: a taxonomy," in *Building the Information Society*, ed: Springer, 2004, pp. 91-120.

[18] Dubrova, E. "Fault tolerant design: An introduction," *Department of Microelectronics and Information Technology, Royal Institute of Technology, Stockholm, Sweden*, 2008.

[19] Mehmood, A; Alrajeh, N; Mukherjee, M; Abdullah, S; Song, H. "A survey on proactive, active and passive fault diagnosis protocols for wsns: network operation perspective," *Sensors*, vol. 18, p. 1787, 2018.

[20] Preparata, FP; Metze, G; Chien, RT. "On the Connection Assignment Problem of Diagnosable Systems," *IEEE Transactions on Electronic Computers*, vol. EC-16, pp. 848-854, 1967.

[21] Barsi, F; Grandoni, F; Maestrini, P. "A theory of diagnosability of digital systems," *IEEE Transactions on Computers*, pp. 585-593, 1976.

[22] Kuhl, JG; Reddy, SM. "Distributed fault-tolerance for large multiprocessor systems," in *Proceedings of the 7th annual symposium on Computer Architecture*, 1980, pp. 23-30.

[23] KUHL, JG. "*Fault diagnosis in computing networks*," Ph.D., Department of Electrical and Computer Engineering, University of Iowa, 1980.

[24] Hakimi, SL; Nakajima, K. "On adaptive system diagnosis," *IEEE Transactions on Computers*, pp. 234-240, 1984.

[25] Hakimi, SL; Schmeichel, EF. "An adaptive algorithm for system level diagnosis," *Journal of Algorithms*, vol. 5, pp. 526-530, 1984.

[26] Bianchini, RP; Buskens, RW. "Implementation of online distributed system-level diagnosis theory," *IEEE Transactions on Computers*, vol. 41, pp. 616-626, 1992.

[27] Duarte, EP; Nanya, T. "A hierarchical adaptive distributed system-level diagnosis algorithm," *IEEE Transactions on computers*, vol. 47, pp. 34-45, 1998.

[28] Duarte, E; Brawerman, A; Albini, LCP. "An algorithm for distributed hierarchical diagnosis of dynamic fault and repair events," in *Proceedings Seventh International Conference on Parallel and Distributed Systems (Cat. No. PR00568)*, 2000, pp. 299-306.

[29] Weber, A; Kutzke, AR; Chessa, S. "Diagnosability evaluation for a system-level diagnosis algorithm for wireless sensor networks," in *The IEEE symposium on Computers and Communications*, 2010, pp. 241-244.

[30] Weber, A; Kutzke, AR; Chessa, S. "Energy-aware test connection assignment for the self-diagnosis of a wireless sensor network," *Journal of the Brazilian Computer Society*, vol. 18, pp. 19-27, 2012.

[31] de Oliveira Barros, M; Weber, A. "System-level diagnosis for WSN: A heuristic," in *2016 17th Latin-American Test Symposium (LATS)*, 2016, pp. 171-176.

[32] Malek, M. "A comparison connection assignment for diagnosis of multiprocessor systems," in *Proceedings of the 7th annual symposium on Computer Architecture*, 1980, pp. 31-36.

[33] Chwa, KY; Hakimi, SL. "Schemes for fault-tolerant computing: A comparison of modularly redundant and t-diagnosable systems," *Information and Control*, vol. 49, pp. 212-238, 6//1981.

[34] Macng, J; Malek, M. "A Comparison Connection Assignment for Self-Diagnosis of Multicomputer Systems," in *Proceedings of the 11th IEEE Fault-Tolerant Computing Symposium*, 1981, pp. 173-175.

[35] Sengupta, A; Dahbura, AT. "On self-diagnosable multiprocessor systems: diagnosis by the comparison approach," *IEEE Transactions on Computers*, vol. 41, pp. 1386-1396, 1992.

[36] Hosseini, SH; Kuhl, JG; Reddy, SM. "A diagnosis algorithm for distributed computing systems with dynamic failure and repair," *IEEE Transactions on Computers*, pp. 223-233, 1984.

[37] Blough, DM; Brown, HW. "The broadcast comparison model for on-line fault diagnosis in multicomputer systems: theory and

implementation," *IEEE Transactions on Computers*, vol. 48, pp. 470-493, 1999.

[38] Chessa, S; Santi, P. "Comparison-based system-level fault diagnosis in ad hoc networks," in *Reliable Distributed Systems, 2001. Proceedings. 20th IEEE Symposium on*, 2001, pp. 257-266.

[39] Elhadef, M; Boukerche, A; Elkadiki, H. "Diagnosing mobile ad-hoc networks: two distributed comparison-based self-diagnosis protocols," in *Proceedings of the 4th ACM international workshop on Mobility management and wireless access*, 2006, pp. 18-27.

[40] Jarrah, H; Chong, P; Sarkar, NI; Gutierrez, J. "A Time-Free Comparison-Based System-Level Fault Diagnostic Model for Highly Dynamic Networks," in *Proceedings of the 11th International Conference on Queueing Theory and Network Applications*, 2016, p. 12.

[41] Dahbura, AT; Sabnani, KK; King, LL. "The comparison approach to multiprocessor fault diagnosis," *IEEE Transactions on Computers*, pp. 373-378, 1987.

[42] Pelc, A. "Undirected graph models for system-level fault diagnosis," *IEEE Transactions on Computers*, vol. 40, pp. 1271-1276, 1991.

[43] Rangarajan, S; Fussell, D. "A probabilistic method for fault diagnosis of multiprocessor systems," in *1988 The Eighteenth International Symposium on Fault-Tolerant Computing. Digest of Papers*, 1988, pp. 278-283.

[44] Marti, S; Giuli, TJ; Lai, K; Baker, M. "Mitigating routing misbehavior in mobile ad hoc networks," in *Proceedings of the 6th annual international conference on Mobile computing and networking*, 2000, pp. 255-265.

[45] Lee, JW; Lee, Y; Syrotiuk, VR. "The performance of a watchdog protocol for wireless network security," *International Journal of Wireless and Mobile Computing*, vol. 2, p. 28, 2007.

[46] Shin, D; Kim, D; Lee, J. "A Reliable Watchdog Protocol with Two-way Mutual Confirmation in Wireless Multi-Hop Networks," in *Reliable and Autonomous Computational Science*, ed: Springer, 2011, pp. 201-221.

[47] Serrat-Olmos, MD; Hernández-Orallo, E; Cano, JC; Calafate, CT; Manzoni, P. "Accurate detection of black holes in MANETs using collaborative bayesian watchdogs," in *2012 IFIP Wireless Days*, 2012, pp. 1-6.

[48] Hernandez-Orallo, E; Olmos, MDS; Cano, JC; Calafate, CT; Manzoni, P. "CoCoWa: A collaborative contact-based watchdog for detecting selfish nodes," *IEEE transactions on mobile computing*, vol. 14, pp. 1162-1175, 2014.

[49] Lal, N; Kumar, S; Saxena, A; Chaurasiya, VK. "Detection of malicious node behaviour via I-watchdog protocol in mobile Ad Hoc network with DSDV routing scheme," *Procedia Computer Science*, vol. 49, pp. 264-273, 2015.

[50] Norihiro, S; Higaki, H. "Cooperative Watchdog in Wireless Ad-Hoc Networks," *DEStech Transactions on Computer Science and Engineering*, 2017.

[51] Kollati, VK. "IBFWA: Integrated Bloom Filter in Watchdog Algorithm for hybrid black hole attack detection in MANET," *Information Security Journal: A Global Perspective*, vol. 26, pp. 49-60, 2017.

[52] Khanna, N; Sachdeva, M. "A comprehensive taxonomy of schemes to detect and mitigate blackhole attack and its variants in MANETs," *Computer Science Review*, vol. 32, pp. 24-44, 2019.

[53] Elliott, C; Heile, B. "Self-organizing, self-healing wireless networks," in *2000 IEEE Aerospace Conference. Proceedings (Cat. No. 00TH8484)*, 2000, pp. 149-156.

[54] Lin, H; Delgado-Frias, JG; Medidi, S. "Using a cache scheme to detect selfish nodes in mobile ad hoc networks," in *Communications, Internet, and Information Technology*, 2007, pp. 61-66.

[55] Harte, S; Rahman, A; Razeeb, K. "Fault tolerance in sensor networks using self-diagnosing sensor nodes," in *The IEEE International Workshop on Intelligent Environments, 2005 (Ref. No. 2005/11059)*, 2005, pp. 7-12.

[56] Zhang, R; Zilic, Z; Radecka, K. "Energy efficient software-based self-test for wireless sensor network nodes," in *24th IEEE VLSI Test Symposium*, 2006, pp. 6, pp.-191.

[57] Marzencki, M; Huang, Y; Kaminska, B. "Context-based collaborative self-test for autonomous wireless sensor networks," in *2011 IEEE 17th International Mixed-Signals, Sensors and Systems Test Workshop*, 2011, pp. 7-12.

[58] Merentitis, A; Kranitis, N; Paschalis, A; Gizopoulos, D. "Low energy online self-test of embedded processors in dependable WSN nodes," *IEEE Transactions on Dependable and Secure Computing*, vol. 9, pp. 86-100, 2011.

[59] Suwatthikul, J. "Fault detection and diagnosis for in-vehicle networks," in *Fault Detection*, ed: IntechOpen, 2010.

[60] Bhoi, SK; Khilar, PM. "Self soft fault detection based routing protocol for vehicular ad hoc network in city environment," *Wireless Networks*, vol. 22, pp. 285-305, 2016.

[61] Zhou, Y; Cheng, N; Lu, N; Shen, XS. "Multi-UAV-aided networks: Aerial-ground cooperative vehicular networking architecture," *IEEE vehicular technology magazine*, vol. 10, pp. 36-44, 2015.

[62] Cai, Z; Fang, Z; Wang, Y. "*Mobile communication device self-testing*," ed: Google Patents, 2018.

[63] Varadarajan, S; Gangadharpalli, S; Golwelkar, U; Rao, KK. "*System and method for self-testing of mobile wireless devices*," ed: Google Patents, 2009.

[64] Panda, M; Khilar, PM. "Distributed soft fault detection algorithm in wireless sensor networks using statistical test," in *2012 2nd IEEE International Conference on Parallel, Distributed and Grid Computing*, 2012, pp. 195-198.

[65] Panda, M; Khilar, PM. "Distributed self fault diagnosis algorithm for large scale wireless sensor networks using modified three sigma edit test," *Ad Hoc Networks*, vol. 25, pp. 170-184, 2015.

[66] Asim, M; Mokhtar, H; Merabti, M. "A self-managing fault management mechanism for wireless sensor networks," *arXiv preprint arXiv:1011.5072*, 2010.

[67] Pasin, M; Fontaine, S; Bouchenak, S. "Failure detection in large scale systems: a survey," in *NOMS Workshops 2008-IEEE Network Operations and Management Symposium Workshops*, 2008, pp. 165-168.

[68] Muhammed, T; Shaikh, RA. "An analysis of fault detection strategies in wireless sensor networks," *Journal of Network and Computer Applications*, vol. 78, pp. 267-287, 2017.

[69] Chen, W; Toueg, S; Aguilera, MK. "On the quality of service of failure detectors," *IEEE Transactions on computers*, vol. 51, pp. 561-580, 2002.

[70] Bertier, M; Marin, O; Sens, P. "Implementation and performance evaluation of an adaptable failure detector," in *Proceedings International Conference on Dependable Systems and Networks*, 2002, pp. 354-363.

[71] Friedman, R; Tcharny, G. "Evaluating failure detection in mobile ad-hoc networks," *Computer Science Department, Technion*, 2003.

[72] Elhadef, M; Boukerche, A. "A gossip-style crash faults detection protocol for wireless ad-hoc and mesh networks," in *2007 IEEE International Performance, Computing, and Communications Conference*, 2007, pp. 600-605.

[73] Zhou, J; Yang, G; Dong, L; Liu, G. "Implementation and performance evaluation of an adaptable failure detector for distributed system," in *2007 International Conference on Computational Intelligence and Security (CIS 2007)*, 2007, pp. 266-270.

[74] Tai, AT; Tso, KS; Sanders, WH. "Cluster-based failure detection service for large-scale ad hoc wireless network applications," in *International Conference on Dependable Systems and Networks, 2004*, 2004, pp. 805-814.

[75] Benkaouha, H; Abdelli, A; Bouyahia, K; Kaloune, Y. "Fdan: Failure detection protocol for mobile ad hoc networks," in *International Conference on Future Generation Communication and Networking*, 2010, pp. 85-94.

[76] Benkaouha, H; Adelli, A; Badache, N; Ben-Othman, J; Mokdad, L. "AFDAN: Accurate failure detection protocol for MANETs," in *2015 International Wireless Communications and Mobile Computing Conference (IWCMC)*, 2015, pp. 733-738.

[77] Benkaouha, H; Abdelli, A; Ben-Othman, J; Mokdad, L. "Towards an efficient failure detection in MANETs," *Wireless Communications and Mobile Computing*, vol. 16, pp. 2939-2955, 2016.

[78] Curiac, DI; Volosencu, C; Pescaru, D; Jurca, L; Doboli, A. "Redundancy and its applications in wireless sensor networks: A survey," *WSEAS Transactions on Computers*, vol. 8, pp. 705-714, 2009.

[79] Chen, J; Kher, S; Somani, A. "Distributed fault detection of wireless sensor networks," in *Proceedings of the 2006 workshop on Dependability issues in wireless ad hoc networks and sensor networks*, 2006, pp. 65-72.

[80] Jiang, P. "A new method for node fault detection in wireless sensor networks," *Sensors*, vol. 9, pp. 1282-1294, 2009.

[81] Razak, S; Furnell, S; Clarke, N; Brooke, P. "Friend-assisted intrusion detection and response mechanisms for mobile ad hoc networks," *Ad Hoc Networks*, vol. 6, pp. 1151-1167, 2008.

[82] Marchang, N; Datta, R. "Collaborative techniques for intrusion detection in mobile ad-hoc networks," *Ad hoc networks*, vol. 6, pp. 508-523, 2008.

[83] Malek, SF; Khorsandi, S. "A cooperative rintrusion detection algorithm based on trusted voting for mobile ad hoc network," in *2013 21st Iranian conference on electrical engineering (ICEE)*, 2013, pp. 1-8.

[84] Wu, JY; Duh, DR; Wang, TY; Chang, LY. "On-line sensor fault detection based on majority voting in wireless sensor networks," in *Proc. of 24th Workshop on Combinatorial Mathematics and Computation Theory (ALGO)*, 2007.

[85] Feng, Z; Fu, JQ; Wang, Y. "Weighted distributed fault detection for wireless sensor networks Based on the distance," in *Proceedings of the 33rd Chinese Control Conference*, 2014, pp. 322-326.

[86] Natu, M; Sethi, AS. "Adaptive fault localization in mobile ad hoc battlefield networks," in *MILCOM 2005-2005 IEEE Military Communications Conference*, 2005, pp. 814-820.

[87] Natu, M; Sethi, AS. "Using temporal correlation for fault localization in dynamically changing networks," *International Journal of Network Management*, vol. 18, pp. 303-316, 2008.

[88] Sapello, A; Sethi, A; Nodine, M; Chadha, R. "Application of time series analysis to fault management in MANETs," in *2010 International Conference on Network and Service Management*, 2010, pp. 150-157.

[89] Nodine, M; Baker, D; Chadha, R; Chiang, CYJ; Moeltner, K; D'Silva, T; et al., "Issues with and approaches to network monitoring and problem remediation in military tactical networks," in *MILCOM 2009-2009 IEEE Military Communications Conference*, 2009, pp. 1-7.

[90] Baker, D; Nodine, M; Chadha, R; Chiang, CJ; Gottlieb, Y; Hsu, CPF. et al., "Computing diagnostic explanations of network faults from monitoring data," in *MILCOM 2008-2008 IEEE Military Communications Conference*, 2008, pp. 1-7.

[91] Gupta, D; Mohapatra, P; Chuah, CN. "Diagnosing failures in wireless networks using fault signatures," in *2010 IEEE International Conference on Communications*, 2010, pp. 1-5.

[92] Vashist, A; Izmailov, R; Manousakis, K; Chadha, R; Chiang, CJ; Serban, C; et al., "Towards network invariant fault diagnosis in MANETs via statistical modeling: The global strength of local weak decisions," in *2012 IEEE Network Operations and Management Symposium*, 2012, pp. 981-987.

[93] Novotny, P; Wolf, AL; Ko, BJ. "Fault localization in MANET-hosted service-based systems," in *2012 IEEE 31st Symposium on Reliable Distributed Systems*, 2012, pp. 243-248.

[94] Novotny, P; Ko, BJ; Wolf, AL. "Locating Faults in MANET-Hosted Software Systems," *IEEE Transactions on Dependable and Secure Computing*, pp. 452-465, 2018.

[95] Karthikeyan, N; Palanisamy, V; Duraiswamy, K. "Performance comparison of broadcasting methods in mobile ad hoc network," *International Journal of Future Generation Communication and Networking*, vol. 2, pp. 47-58, 2009.

[96] Jüttner, A; Magi, Á. "Tree based broadcast in ad hoc networks," *Mobile Networks and Applications*, vol. 10, pp. 753-762, 2005.

[97] Elhadef, M; Boukerche, A; Elkadiki, H. "A distributed fault identification protocol for wireless and mobile ad hoc networks," *Journal of parallel and distributed computing*, vol. 68, pp. 321-335, 2008.

[98] Elhadef, M; Boukerche, A; Elkadiki, H. "Self-diagnosing wireless mesh and ad-hoc networks using an adaptable comparison-based approach," in *The Second International Conference on Availability, Reliability and Security (ARES'07)*, 2007, pp. 983-990.

[99] Ji, L; Xu, L. "A two-level scheme for fault diagnosis based on the comparison model in Wireless Mesh Networks," in *2011 IEEE 13th International Conference on Communication Technology*, 2011, pp. 592-596.

[100] Sahoo, MN; Khilar, PM. "System-level fault diagnosis in fixed topology mobile ad hoc networks," *International Journal of Communication Networks and Distributed Systems*, vol. 10, pp. 216-232, 2013.

[101] Aljeri, N; Almulla, M; Boukerche, A. "An efficient fault detection and diagnosis protocolfor vehicular networks," in *Proceedings of the third ACM international symposium on Design and analysis of intelligent vehicular networks and applications*, 2013, pp. 23-30.

[102] Sahoo, MN; Khilar, PM. "Intermittent fault diagnosis in dynamic topology MANETs," *International Journal of Signal and Imaging Systems Engineering*, vol. 8, pp. 345-355, 2015.

[103] Yu, JY; Chong, PHJ. "A survey of clustering schemes for mobile ad hoc networks," *IEEE Communications Surveys & Tutorials*, vol. 7, pp. 32-48, 2005.

[104] Li, D. "Cluster-based system-level fault diagnosis in hierarchical ad-hoc networks," in *2007 International Conference on Computational Intelligence and Security (CIS 2007)*, 2007, pp. 1062-1066.

[105] Chiang, CC; Gerla, M. "Routing and multicast in multihop, mobile wireless networks," in *Proceedings of ICUPC 97-6th International Conference on Universal Personal Communications*, 1997, pp. 546-551.

[106] Yadav, N; Khilar, P. "An improved hierarchically adaptive distributed fault diagnosis in mobile ad hoc networks using clustering," in *2010 First International Conference on Integrated Intelligent Computing*, 2010, pp. 308-313.

[107] Hutle, M. "An efficient failure detector for sparsely connected networks," in *Parallel and Distributed Computing and Networks*, 2004, pp. 369-374.

[108] Hassan, R; Jarrah, H. "Comparison Approach System-Level Fault Diagnosis in MANET Using Non-Overlapping Clustering Technique," presented at the *Mosharaka International Conference on Wireless Communications and Mobile Computing* Istanbul, Turkey, 2011.

[109] Yu, J; Chong, PH. "3hbac (3-hop between adjacent clusterheads): a novel non-overlapping clustering algorithm for mobile ad hoc networks," in *2003 IEEE Pacific Rim Conference on Communications Computers and Signal Processing (PACRIM 2003)(Cat. No. 03CH37490)*, 2003, pp. 318-321.

[110] Konstantopoulos, C; Gavalas, D; Pantziou, G. "Clustering in mobile ad hoc networks through neighborhood stability-based mobility prediction," *Computer Networks*, vol. 52, pp. 1797-1824, 2008.

[111] Ruiz, P; Bouvry, P. "Survey on broadcast algorithms for mobile ad hoc networks," *ACM computing surveys (CSUR)*, vol. 48, p. 8, 2015.

[112] Wang, J; Xie, B; Agrawal, DP. "Journey from mobile ad hoc networks to wireless mesh networks," in *Guide to wireless mesh networks*, ed: Springer, 2009, pp. 1-30.
[113] Jarrah, H; Chong, PH; Sarkar, NI; Gutierrez, J. "Efficient Fault Identification Protocol for Dynamic Topology Networks Using Network Coding," in *International Conference on Smart Grid Inspired Future Technologies*, 2018, pp. 230-239.
[114] Lun, DS; Médard, M; Koetter, R; Effros, M. "On coding for reliable communication over packet networks," *Physical Communication*, vol. 1, pp. 3-20, 2008.

In: Vehicular Ad-Hoc Networks ISBN: 978-1-53618-038-1
Editors: G. G. Md. Nawaz Ali et al.

Chapter 5

BLOCKCHAIN: BASICS, APPLICATIONS, CHALLENGES AND OPPORTUNITIES

Jaswant Arya[1], Arun Kumar[2, *], Akhilendra Pratap Singh[1], Tapas Kumar Mishra[2] and Peter H J Chong[3]

[1]Department of Computer Science and Engineering,
National Institute of Technology, Meghalaya, India
[2]Department of Computer Science and Engineering,
National Institute of Technology, Rourkela, India
[3]Department of Electrical and Electronic Engineering,
Auckland University of Technology, New Zealand

ABSTRACT

Today, centralized systems are more common, and the reversibility of the process and transparency are kept hidden due to the ownership of a single authority. The consistency is difficult to maintain as the network is centralized, and the central server is responsible for managing overall traffic. Blockchain brings out the new features like decentralization,

[*] Corresponding Author's Email: kumararun@nitrkl.ac.in.

distribution, immutable, consistent and security by cryptographer hash algorithms. Blockchain technology is based on the consensus which hires distributed ledgers for enabling parties who do not fully trust each other to maintain a set of global states. The intermediaries work to maintain trust and the reduced number of intermediaries can make the process easier, faster, transparent and cheaper. The consensus mechanism is used to validate the transactions, and privacy is maintained by having multiple account addresses. This chapter presents the basics of Blockchain technology, applications, research challenges and opportunities in the field. As a recent phenomenon, Blockchain has the potential to bring a new perspective to security, resiliency and efficiency in the digital world. This chapter discusses each issue in details.

Keywords: blockchain, cryptocurrencies, architecture, consensus algorithms, smart contract, bitcoin, applications

1. Introduction

Blockchain is a decentralized, distributed, and immutable ledger secured by cryptographic hash algorithms. Blockchain ledger has chained chronological and encrypted blocks of the synchronized data across the peer to peer network. The data is formed into blocks, and each block is appended to its previous block connected by hash. The blockchain first stores the data into back linked blocks. It then verifies the blocks with a distributed consensus process to keep the security, privacy, and transparency in the whole blockchain network. The most prominent and widespread application of blockchain technology is Bitcoin, the first cryptocurrency [1].

Blockchain has become the most promising technology for decentralized trust without any third party. The data-driven autonomous industries adopted the blockchain technology as the mechanism to store the data immutably in a distributed and secured manner. The digital tokens used in blockchain for transferring and accessing the digital assets between peer to peer users proves its prominent use to remove the need of any third party [2]. Data security, as well as data integrity, has been achieved by cryptocurrency and smart contract [3]. For executing the program on different machines in a distributed manner with the consensus process, the blockchain can be applied as an additional layer [4]. It is one of the reasons why blockchain has given birth cryptocurrency and smart contracts that impose the critical requirement on data security and integrity [5]. For general bytecode execution, the orchestration for the global state can be achieved with the distributed consensus and Blockchain technology is expected to evolve the open access and transaction management on decentralized autonomous systems [6]. It involves the decentralization of the assets in the finance sector, IoT, in public services and decentralized applications, etc.

Each new transaction is validated and recorded in a global ledger by a miner. Each miner contributes his computational power to verify the data blocks of distributed ledger and can compete in a consensus process to create the next new block. Once the block is created, it can be appended to the main chain. In the blockchain, data is of two types- on-chain data which is to be kept in blockchain and off-chain, which can not be stored in the block so that the verification process may be less complicated [7]. Blockchain uses a decentralized token system to authenticate the asset transfer. Furthermore, a Blockchain maintains an arbitrary order of the transactional records by cryptographically chaining the record subsets in the form of data "blocks" to their chronic predecessors. Data tampering can be prevented by the linked block using cryptography. Consensus protocols maintain the consistency of the transaction in the weak synchronized network. The authors [8] have reviewed the blockchain use cases and its adaptability in IoT for integrity, anonymity, and the interoperability. The authors in [9] have presented the blockchain as a solution for security, privacy, and computational limitations of different IoT devices.

The details of the transfer are recorded on a public ledger that anyone on the network can access. Blockchain is similar to the distributed public ledger, in which all transactions are stored in a chain of blocks. So this chain continuously grows when new blocks are appended to it. Blockchain technology has essential characteristics, such as decentralization, persistence, anonymity, and auditability. Blockchain can work in a decentralized environment, which is enabled by integrating several core technologies such as cryptographic hash [14], digital signature [15], and distributed consensus mechanism. Blockchain has provided a much powerful security alternative when compared to database systems [17]. In a decentralized transaction, blockchain technology can reduce the cost and improve efficiency without the existence of any third party.

Blockchain technology has shown its potential by its main characteristics decentralization, distribution, and persistence. The cryptographic hash algorithms contribute more in the security and digital signature to add encryption between peer to peer user. Below, we list a few of the properties of the blockchain.

1. *Decentralization*: In blockchain technology, the data is decentralized so that each node, which is the part of the blockchain network has consistent data. A central, trusted authority authorizes centralized data. In most cases, this centralized authority works well, but the increasing number of intermediaries trigger the cost of the transaction. For example, in the banking system, the banks are centralized trusted authorities, and in most cases, it works well. However, this mediation limits the minimum practical transaction cost and cuts its incentive. In a blockchain network, the transaction can be completed without a centralized authority. This helps in reducing the overall cost when multiple transactions are used for the same product. Also, the workload of the central server can be minimized significantly by partitioning the verification on multiple nodes.
2. *Persistency:* In the blockchain network, the transactions are inherently stored in blocks for creating a chronologically decentralized and distributed chain which is pursued by the entire network. Each node contains the blockchain ledger to recover from any data loss by a node. Persistence and consistency are easily achievable by storing the data in different nodes and detecting the missing data or any falsification.
3. *Anonymity:* In the banking system, the identity is stored by the banks to verify the transactions, and the personal details of the

transaction can easily be traced. However, in the blockchain network, a person interacts with the generated wallet address and personal identity is kept hidden by using multiple addresses and there is no central authority which stores the private data of a node. Transactions are verified publicly, and the privacy is retained by using the addresses for each transaction. This ensures that a miner cannot know the identify the owner or the issuer of the wallet address. Also, a central party does not ask for the private data of a user.

4. *Transparency:* Blockchain network provides transparency for the whole network by enabling distributed and consistent transaction records. In the supply chain and logistics industry, the requirement of transparency cannot be ignored. Blockchain network reduces the number of frauds by establishing a transparent system.
5. *Auditability:* The blockchain network includes the timestamp that can be used to trace the transaction detail, and the user can verify the transaction easily. The higher auditability is proved by the transparency and traceability of the transactions.
6. *Time reduction:* Each software in the blockchain network has consistent information and therefore removing the lengthy process of verification. The only single version of agreed can reduce the time by avoiding the need for a lengthy process of verification, settlement, and clearance.

Earlier research work has discussed the blockchain in IoT, blockchain in healthcare, blockchain in the logistics industry, and a few have discussed blockchain consensus process and architecture. The authors [18] presents a smart home application for blockchain in IoT. The authors [19] describes the use of a smart contract to create a business model. The authors [20] reviewed the integration of blockchain with IoT and the use cases of this integration. The work [21] raised the security issues in the IoT and brought the advantages of blockchain to resolve these issues. Many surveys are focusing on the integration of blockchain with the specific area or limited field. Still, there is no universal and general survey in the best of our knowledge which combines the architecture of blockchain with the applications and the challenges.

This chapter presents the underlying architecture, mechanism, applications, challenges, and opportunities of the blockchain technology. Moreover, the key characteristics of blockchain, such as decentralized, immutable, distributed, and secured, are discussed. The comparison between different versions and variants of the blockchain technology as well as the fundamental difference between the various consensus algorithms are presented in details. Also, the basic principle of cryptocurrencies and their applications are presented.

2. Blockchain Architecture

Blockchain is the chain of sequentially connected blocks holding the complete list of transaction records. Except for the first block, every block has a parent block. The blocks can be reached by traversing back as every block contains the hash of its parent block. The first block is known as the genesis block. The block number identifies each block, and each block contains valid transactions. The transactions are not stored directly, each transaction is hashed, and Merkle tree of these hashes is formed. The timestamp is always attached with a transaction for maintaining the chronological order.

A. Block

A block is the data structure which contains a hash of the parent block, transactions, Merkle root, timestamp, etc. The first block of a blockchain is called Genesis Block. Block version indicates the set of block validation rules to follow. The previous block of any block is called Parent Block, and the Hash of the parent block is called parent Hash.

B. Hash of the Block

A hash can be understood as a fingerprint, which is unique for each block. It identifies a block and all of its content, and it is always unique. Therefore, once a block is created, any change inside the block will cause the hash to change. In the following example, the value returned by a hash is a fixed-length string which contains only two things -the upper case alphabet and digits. Any complex function h(x) can be used to generate the hash. For making cryptographic hash secured, the highly complex hash algorithm like SHA-256 can be used [11]. For instance, MD2 hash produces the following results-

Input: jaswant
Hash Output:d2ebbf629b8c2d492321eda2e6d3ef6b
Input: jaswanu
Hash Output: 964910d763b8ffbd35631e7b74a375f1

The hash function is created in such a way that a single change in data leads to a drastic shift in output hash value. It is impossible to assume the input data based on the hash value. Also, more than one input data cannot give the same hash value.

C. Time Stamp

A timestamp server can prevent the double-spending problem. The hash value of the block is published in blockchain network similar to news publishing in a newspaper. Timestamp can prove the validation of data at the right time to get into the hash. The chain of hash is formed by adding the timestamp reinforcing the ones before it.

D. Difficulty

The difficulty is the value which decides the difficulty level to calculate a hash threshold for a given target. If the number of transactions increases, then it is challenging to create the new blocks in high speed and of course there may be chances of evolving number due to attackers or greedy miners. Therefore, to control the speed, the difficulty level is increased—the level of difficulty increases as the speed of block formation increase.

E. Nonce

Nonce stands for a pseudo-random number used only once during the process of mining. In Bitcoin blockchain, a nonce is a number proceeding a certain number of zeros. During the mining process, the nonce is found spending the energy to process the algorithms. Thus replay attack can be exhibited in block creation.

F. Block Header

A block header is a string which contains block version, previous block hash, Merkle tree root hash, timestamp, nBits difficulty, and Nonce. A block header contains transactions as the information to be secured. A block can have multiple transactions depending on its size. In Bitcoin blockchain, each block size is 1 MB [1].

G. Blockchain Data Structure

The Blockchain data structure is an ordered back-linked list of blocks of information/transaction/data. It can be created by forming a flat file or a simple database. The identity of the block is kept by a hash which is calculated from cryptographic hash algorithms such as SHA512, SHA256, etc. Every block contains the parent hash. If an attacker tries to make changes in the block, the hash of that block will be changed, and then the proceeding blocks will not be further connected to this block due to mismatched parent hash. If a new block is created, it is appended at the end of the blockchain. The process of creating a genesis block is different as there is no block on which it is appended. Therefore the initialization is done to form a genesis block with zero parent hash.

Figure 1 illustrates the block structure of one block in the Blockchain having one parent hash or previous block hash, one difficulty level, one nonce value, multiple transactions and Merkle root formed by the hashes of transactions.

In the Merkle tree, the hash of the parent node is calculated by appending the hash of the right node to the left node and then hashing to generate the hash of the parent node. All the underlying transactions are combined and hashed together to form the Merkle tree.

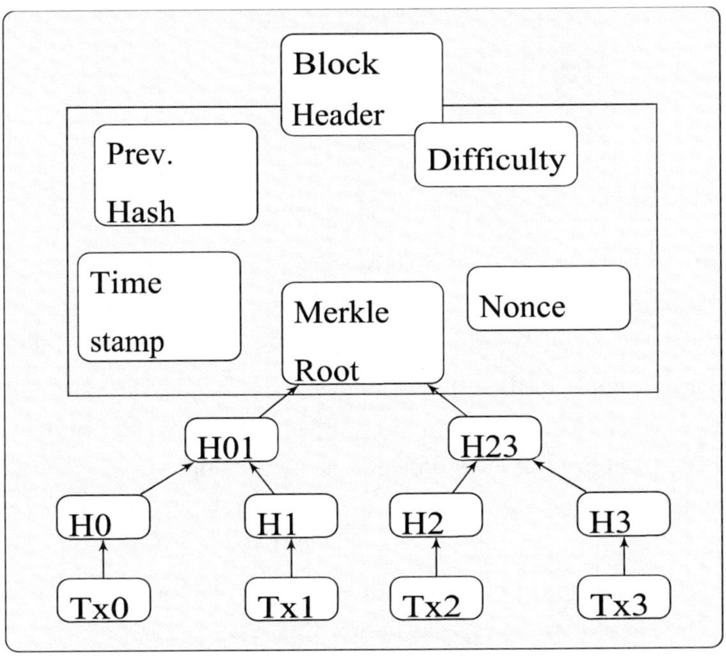

Figure 1. Blockchain block structure [1].

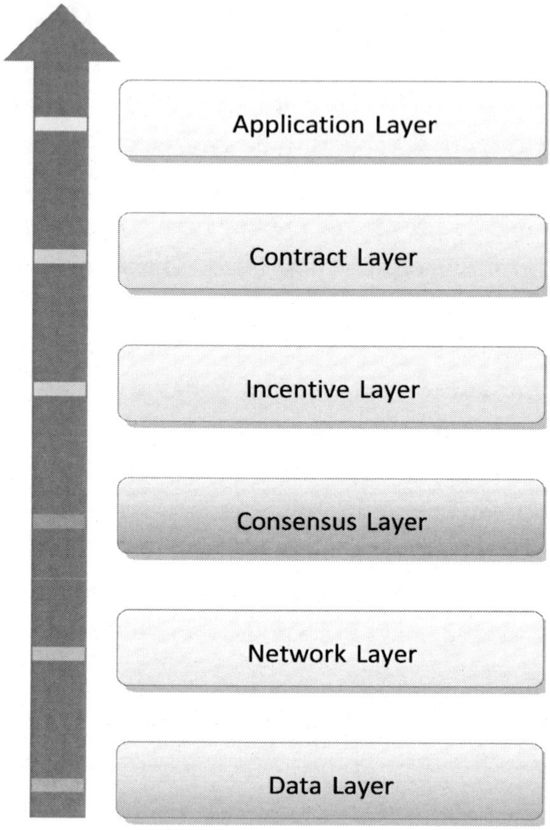

Figure 2. Blockchain layers [45].

Figure 2 illustrates the six different layers of the blockchain network. The blockchain network has different layers to perform specific tasks similar to the ISO reference model of the Internet [45], [46]. Data layer collects the data and stores it into a ledger using the asymmetric encryption techniques, hashing, time-stamped Merkle trees. The network layer provides the functionality to connect all the nodes in the network and creates a decentralized peer to peer network environment. The consensus layer maintains the consensus via the different consensus algorithms among the network nodes and maintains the consistency of the transaction records. The incentive layer distributes the reward to the miners on the successfully mining a block. Contract layer contains the smart contract to manage the data of money or assets records etc. Finally, the application layer provides different applications and use cases of blockchain.

H. Digital Signature

The encryption techniques are used to authenticate, integrate a message to keep its confidentiality and privacy. The digital signatures are more like a stamped seal or handwritten signature, and thus, it solves the challenges of tampering and fake authentication or impersonation in communication [15]. The elliptic curve digital signature algorithm (ECDSA) [12] is widely used for Blockchain network. In the blockchain network, the peer to peer connection is established between all the nodes, and each node owns the consistent ledger. This consistent ledger is the proof of a transaction. This ledger is in the form of sequentially chained blocks connected with its previous block by a hash.

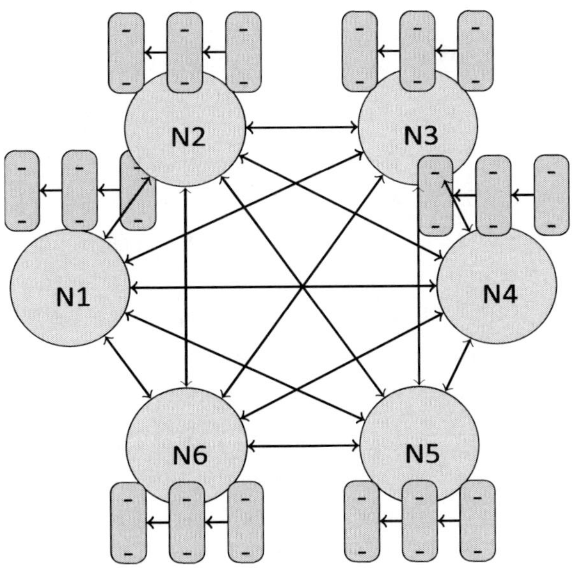

Figure 3. Illustration of the peer to peer connected nodes in an open-access Blockchain network.

Every block consists of the hash value of the previous block similar to the singly linked list. The new transactions are kept in these blocks. Figure 3 illustrates the established peer to peer network between six nodes ranging from *N1* to *N6* with each node having its ledger to verify the consistency of transaction records. The higher security is achieved by sequential and unidirectional traversing connectivity of the blocks.

3. Consensus Algorithms

The blockchain network must confirm the consistency and clarity of the ledger in each software. This is completed by a few protocols which proceed the transaction process having absolute standards. The verification process is continued in a decentralized manner as performing the transactions on the distributed database managed by peer to peer connected nodes in the blockchain network. After validating and verifying the block, the block of the transaction is published in the network. While it might be possible that more than one miner is creating the block simultaneously, and this is a challenge to maintain the blockchain network. The various protocols are used to avoid inconsistency. The techniques or algorithms which are used to reach the consensus are called the consensus algorithm. The different strategies are proposed to reach the consensus like proof of work, proof of stake, proof of authority, and Practical Byzantine Fault Tolerance (PBFT). Each consensus algorithm has its strategy to pass the agreement and may use different protocols. This paper describes in details the backbone of the consensus algorithms.

The process of consensus is activated by requesting the first transaction to trigger the creation of a new block. After the mining process of the new block holding the transaction, this mined block is broadcasted to every node. One can accept the valid block and also can reject the invalid block. In Figure. 4, the flow chart illustrates the simplified process of transaction requests, block creation, and its circulation within the blockchain network.

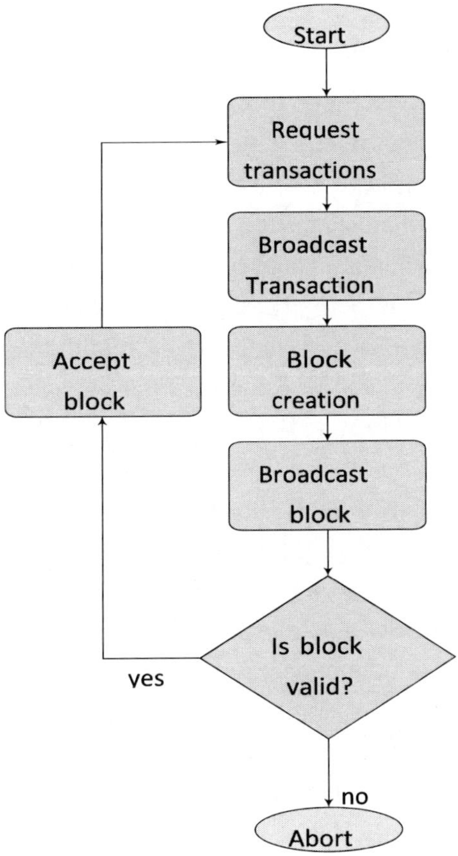

Figure 4. Overview of the Blockchain process.

A. Proof of Work (PoW)

In proof of work consensus process, every node which wants to participate in the consensus process has to contribute its resources to solve the mathematical puzzle. This puzzle has a different level of difficulties. The transactions are broadcasted into the network to be added into the memory pool by the active nodes. The target to find a value called Nonce. Nonce has few sets of rules like the generated value must have a fixed number of zeros as a prefix. When a miner finds the required Nonce, he/she gets the right to create the new block and store the transaction into the block according to the decreasing order of the transaction fees. Then this block is broadcasted into the blockchain network. Other nodes can accept or reject after verifying the transactions of the block. They can accept the block by appending it to their chain. The consensus process starts with the competition to take a right to pack the new transactions and continues until the block acceptance.

Proof of work mechanism is applied to distribute the time stamp server to the entire blockchain network. In this mechanism, a hash value is calculated having *n* number of zero bits by scanning the Nonce. The Nonce is incremented until its scanned value puts the required number of bits for the block hash. The CPU effort is spent on every calculation of this work. This calculation requires the work exponential in the number of bits required. Each block is uniquely linked to its previous block except the genesis block and hence redoing the work for any block directs to redo the work for subsequent blocks. Therefore it is challenging to change any block and its subsequent block. Proof of work consensus mechanism is based on one CPU, one vote concept to verify the transaction and to accept the longest chain. The exact and accurate chain can only be created by an honest node and form the longest chain. If an attacker alters the transaction in a single block by redoing the proof of work, he has to redo the proof of work for each subsequently linked block. The hardware speed is controlled by creating the difficulty as a target value to perform proof of work. If blocks are generated fast, this difficulty target increases to compensate the hardware speed and hence, a specific increasing time is consumed to acquire the proof of work. An attacker can gain the 51% power of the network, and this scenario ends up the blockchain use. The authors in [13] proposed the hybrid consensus mechanism of proof of work and proof of stake.

B. Proof of Stake (PoS)

Proof of work allows every node to participate in the mining of the block, but among all those miners, only one miner receives the right to create the new block and all other nodes waste the incentive and the energy of the network. Moreover, if 51% of the miner attain the potential to mine faster than others by having the higher energy to mining compared to other miners, he may rule the whole blockchain network. Therefore the proof of stake works better by allowing only all those nodes who prove themselves the stakeholder. It saves the condition of one node owning the network because only one node can not have 51% money of the whole network. There is a probability of owning the network by one node in the beginning, but later on, it becomes almost impossible to have money more than the whole network. The proof of stake can efficiently reduce the energy consumption by proof of work because the richer are not supposed to attack the network. Thus the transaction speed can be boosted up by reducing the number of miners. If the wealthiest person is dishonest, and he/she tries to own the whole blockchain network, he has to own more than half of the total currency of the network [63].

C. Practical Byzantine Fault Tolerance (PBFT)

The process to create consensus among byzantine soldiers is more difficult if there are no nodes which can begin the process to reach on an agreement. The possible byzantine fault tolerance (PBFT) is based on the higher weight of the leader, so that final agreement to be without any confusion. The byzantine faults are to be solved by pdf. In PBFT the consensus is reached in multiple rounds.

In the first round, a new block is created, and in the subsequent rounds, the primary node is chosen based on predefined protocols. These multiple rounds can reorder the transactions. Proof of work requires the hash process to reach the consensus, but in PBFT, a consensus is more likely to be reached using the voting process.

A node can reach to the preparation phase from preprepared phase if it receives 2/3 rd nodes favor of all the nodes of the network. Each node has to receive 2/3 rd favor nodes to enter in the commit phase from the preparation phase again. Thus the consensus process is completed in three subprocesses. The node should be active while voting to receive the agreement from each node. The nodes can ask their neighbor nodes for any information for such as voting process or consensus process. The weight of votes for a few nodes might be higher or lower depending on the network node types. Hyperledger Fabric uses PBFT consensus for creating its product based on blockchain technology [64].

D. Delegated Proof-of-Stake (DPoS)

For energy-efficient process, the number of nodes which participate in the mining process can be decreased by allowing the nodes which get priority according to their stake. Unlike to PoS, DPoS reduce the number of verifiers by awarding the priority to the top stakeholder and make the consensus process faster and energy-efficient. This priority rule names the DPoS as the representative democratic process. In Bitshares, the DPoS is used as the consensus process. EOS[27] is based on delegated proof of stake.

E. Proof of Space (PoS)

Disk space can be another choice to select the miners because of the more disk space, the more probability to mine a block successfully. By initializing the dedicated disk space to mine, the block can be energy efficient. The unused disk space can be reused by the miners to make the miner process cheaper.

If a miner node mine the block unsuccessfully then he would not lose the much higher energy like proof of work. Thus, Proof of space has a very high economic advantage, and every node always can invest unused disk space [16].

F. Proof of Authority

If a node creates the block, this block would be verified by the authenticated nodes in multiple and different scheduled rounds. In a different schedule, the authenticated nodes are asked to propose the next block [10].

G. Ripple

If one of the subnetworks is trusted, its nodes can be listed separately and be asked to verify the transactions. Unlike PBFT, Ripple uses the nodes of intersecting groups with the most of non-Byzantine nodes. Server node can verify the transactions and create the block. Client nodes can transfer the funds. There is a different node list to be queried by the server when the transactions are stored into the ledger [30].

H. Tendermint

In this consensus process, the block is created and confirmed in three steps: prevote, pre-commit, and commit. In the first step, a prevote is broadcasted for the newly created block as a proposal. In the second step, a pre-commit is broadcasted if the prevote is favored by 2/3 of all the nodes. At last, the commit is broadcasted if pre-commit is favored by 2/3 of all the nodes and block is validated. A node can accept the block if the commit is favored by 2/3 of all the blocks [34].

Table 1 presents the comparison between different consensus algorithms. This comparison is derived from [63]. The blockchain network can follow any consensus algorithm to validate the block and to maintain the network established with honest nodes. If an attacker tries to disrupt the network, he should be identified and punished by the honest nodes, and therefore, the private industry can create its network of nodes to be secured and be more auditable.

Table 1. Comparison between different consensus algorithms

Consensus Algorithm	Implementation Purpose	Description	Simulated Random Function	Puzzle Design	Practical Example
Proof of Work	Sybil attack [54]	Calculating hash repeatedly	Used	Single task	Bitcoin and Litcoin [55]
Proof of Retrievably	Distributed Storage	No interaction using Fiat Shamir transformation and random Merkle proofs	Can be used	Two-step task	Permacoin [56], Kopper Coin [57]
Proof of space [16]	Decentralised Storage market	Redoing the PoR over time	May be used or may not be	Two-step process	Filecoin [58]
Equihash [61]	ASIC resistance	Time complexity trade-off in proof generation	Used	Memory hard	Zcash
Ethash [60]	ASIC resistance	Calculation of cryptographic hash	Used	Sequential and memory-hard algorithm	Ethereum [26]
Proof of space	Easier participation in Consensus	Random Merkle proof	Used	Two-step process and measurement of proof quality	Space Mint

To save the disk space, the transactions contained by the blocks are discarded, but the block hash remains unchanged. If the root hash is achieved, then the branch hashes are stubbed off from the old blocks. Initially, there is no issue to store the blockchain of a smaller number of transaction. But, in future, the increasing number of transactions may require a higher speed of block-creation and the massive disk space. The complexity of block creation deduces the wastage of time and may trigger the situation of insufficient disk space. The older transactions might not be required, and then they must be discarded to save the disk space as well as the complexity of blockchain [1]. Figure 5 shows the Blockchain Block Structure consists of Merkle tree hashes.

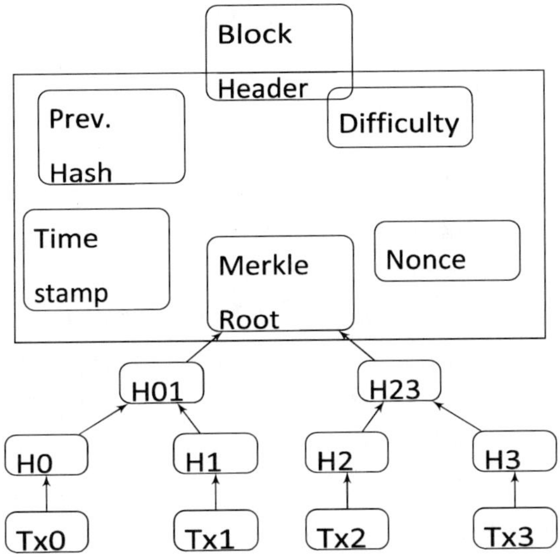

Figure 5. Blockchain block structure consists of Merkle tree hashes [1].

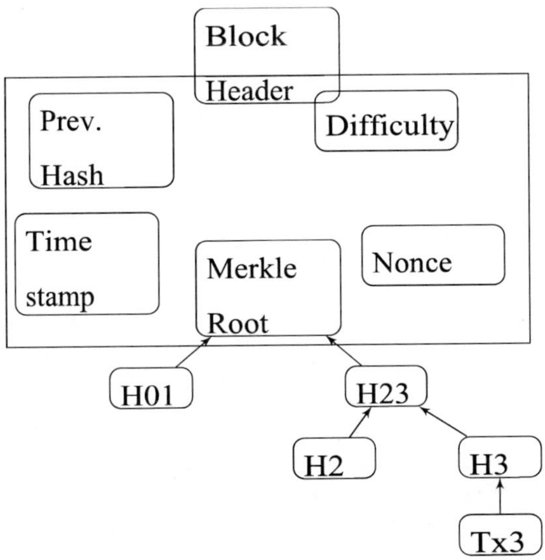

Figure 6. Blockchain block structure after stubbing off the branches to diminish transaction 0, 1 and 2 [1].

Figure 6 illustrates the block having discarded branches of the Merkle tree for the older transaction. Forking is possible in blockchain when a node accepts two or more blocks simultaneously. Each miner is independent of other miners to create and validate the block, and it may be possible that more than two miners generate the block simultaneously. In this case, a node can accept both the blocks and the chain can grow for both the blocks until one of them proves to be authentic and to be retained. The more massive chain is retained. It takes a specific time to generate and broadcast the newly mined block. We observe a case when two miners generate the blocks simultaneously and broadcast it. In that case, a node can accept both the blocks to create the fork with accepting further blocks.

Figure 7 illustrates the forking condition when two blocks (x+3) and (x+3)' are mined simultaneously. In this scenario, the nodes can accept block either one of them x+3 or x+3' or both. This chain grows until one longest chain is not perceived. Both the chains grow until the one chain becomes longer than other branches, and then the longer one can be accepted by every node. Figure 8 shows an example of the acceptance of the longest chain.

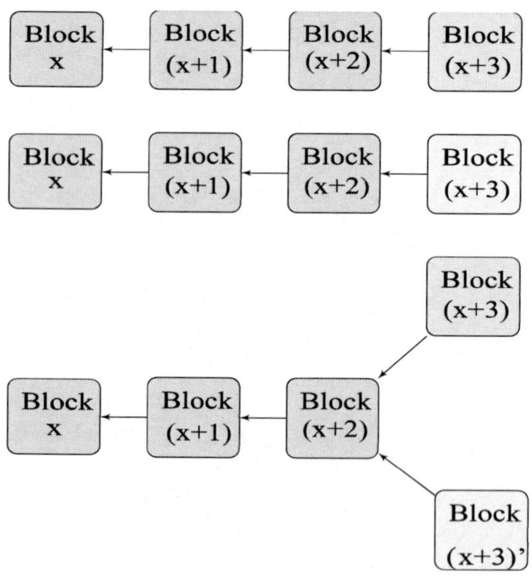

Figure 7. Blockchain forking.

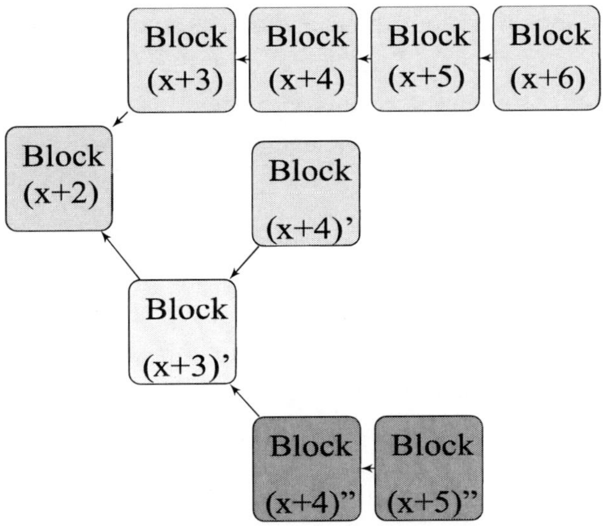

Figure 8. Acceptance of longest chain.

4. Blockchain Variants

The participation of nodes in blockchain network may vary depending on the permission to take part in the consensus process. The set of nodes or a single node should be permitted. The read permission may differ depending on the individual nodes. There are three variants of the blockchain network.

A. Public Blockchain

In a public blockchain network, everyone can verify, broadcast, and add the block to the blockchain ledger; it means that everyone can take part in the consensus process. A computer having internet can join the network as a node. The record of transactions is secured in multiple nodes of the blockchain network. Anonymity is maintained by the complicated generating address for each user.

Redundancy of the blockchain makes it secured against the massive attacks, but it also increases the latency of the process and causes the additional wastage of the electricity. For creating a transparent, secured decentralized network, the public blockchain is preferred most.

B. Private Blockchain

The blockchain platform can be used either as a general-purpose software or a specific purpose software. Ethereum has provided the global platform to create any decentralized blockchain application. Ethereum tools are used to create smart contracts and decentralized applications [35]. Hyperledger has also provided its tools to create more reliable blockchain decentralized application. Not limited to a smart contract, it provides more functionality for a vast area of applications such as the finance, supply chain, and logistics, health-care, and traffic signal systems, etc. [36]. It allows the private and the consortium blockchain tools which can be used by various individual organizations.

There are other platforms like Hijro [37], Tierion [38], Gem [39] and Provenance [40] which can be used for a specific purpose. The company can use the public or restricted platform according to its requirement as Provenance or Hijro for logistic, Tierion or Gem for healthcare, etc.

5. Blockchain Versions

There are many applications based on blockchain technology. All these applications come under three categories Blockchain 1.0 or cryptocurrency, Blockchain 2.0 or smart contract and Blockchain 3.0 or decentralized autonomous organizations (Dapp). The first blockchain application, Bitcoin, started the concept of cryptocurrency in the financial sector. After a few years, the new version of the blockchain application came as a smart contract. The cryptocurrency and smart contract are successful, but the concept of Dapp has generated multiple number of applications in the public sector [42]. Figure 9 illustrates the different versions of Blockchain and their applications/Implementations.

A. Blockchain 1.0

Blockchain 1.0 version can be understood as a database that has the record of transactions in the group of linked blocks. Each block is connected with its previous block hash in the chronological order. There is a distributed consensus mechanism to verify, validate, and append the block in the transaction ledger. The cryptographic hash functions and peer to peer connection are used to obtain higher security. The transaction ledger is stored in each node. There is a public key and a private key. The sender can access its account by the public and private key and can send the message to other peers by knowing his public key. The message can not be modified in anyway.

B. Blockchain 2.0

After the success of cryptocurrencies, another version of the blockchain smart contract brought the next phase, which offers the digital contracts based on the blockchain technology. This type of blockchain provides the decentralized self-executable programs which are executed when certain predefined conditions are satisfied.

This type of blockchain is used to create smart contracts for different purpose like ownership of the vehicle, home or the things, etc. The scripts can be written in the solidity, Go or any other suitable language for defining the conditions and rules of the contract. These conditions are stored in the blockchain ledger [41].

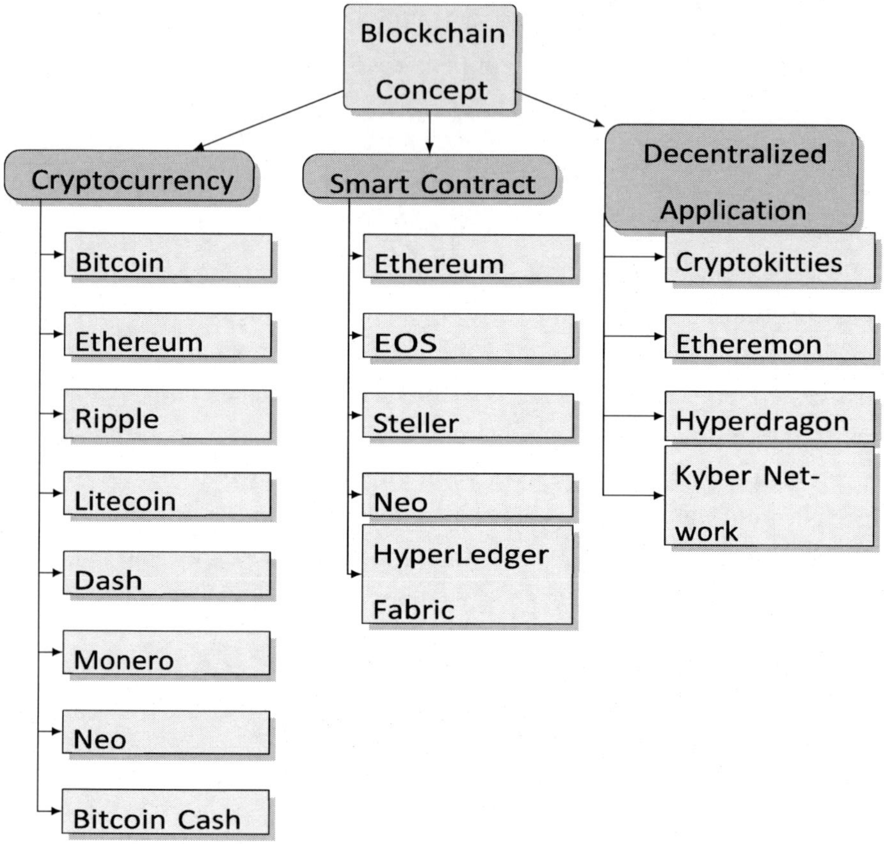

Figure 9. Different version of Blockchain.

C. Blockchain 3.0

Blockchain 3.0 or decentralized autonomous organization offers the implementation of the smart contract. In the public sector, decentralized applications can promise a vital role. Dapp is an abbreviation of the blockchain. Blockchain is trusted more than any other technology, and therefore, Dapp codes are made as an open-source to get the audits from third parties. The tokens are too used to value the support of participants of the system. All the components of the Dapp are hosted and executed in a decentralized manner [43].

Table 2. Different cryptocurrencies

Cryptocurrency	Objective
Bitcoin Cash [50]	Time reduction by increasing the block size
Zilliqa [51]	Higher security by using the more powerful cryptographic techniques zero-knowledge proof
Ethereum	Enhance the programmability of the Blockchain network, smart contracts and Dapps
USDT	maintain the stability between the national currency and cryptocurrency
Peercoin	innovative consensus mechanism
EOS	innovative consensus mechanism
IOTA	created for the Internet of things
Ripple	created for global financial settlement
Augur	prediction of market application

6. CRYPTOCURRENCY

Cryptocurrency is highly secured, distributed, decentralized, and peer to peer digital currency, which is based on Blockchain technology. It requires distributed verification of transactions without a central authority. Verification or validating of the transaction includes confirming transaction amounts, storing transaction history and preventing from double-spending [48]. Cryptocurrency system provides the user with a wallet with a generated address as a public key with a private key to sign transaction as an owner. In some cases, a proof-of-work or proof-of-stake or any other consensus scheme is used to create and manage the currency. The maximum size of the block is fixed. A block of cryptocurrency blockchain contains block header, block version, timestamp, a hash of the previous block, nbits, nonce, Merkle root, hash, difficulty, and transaction counter for storing the transactions. The computer or node which validates and creates a new block is called validator or miner. Generally, for validating and creating the block, one computer has to spend energy, electricity, time, and more space in the existing memory to contain a full copy of all the connected blocks. Till 8th July 2019, there is more than 2200 type of cryptocurrencies in the market with the current market cap of more than $330 billion. Among all the cryptocurrencies, Bitcoin has the highest market cap around $210 billion. Few of them are listed in Table 2 in the decreasing order of their market cap [47]. Centralized service providers or government-controlled the earlier digital currencies, but Cryptocurrencies are entirely decentralized and controlled by distributed consensus algorithms among the different nodes of the peer to peer network. The identity of a person is kept hidden by using the address instead of a user name or any other data. In the centralized system, the user information is always preserved by the service providers.

A. Bitcoin

The traditional electronic cash systems do not use the address, but as an email, Bitcoin uses the address to hide the identity. Bitcoin network can circulate a maximum of 21 million Bitcoin to avoid inflation or deflation. Unlike traditional digital currency, Bitcoin source codes are open, and anyone can see and understand how Bitcoins are generated.

The value of Bitcoin is directly proportional to its active users in the network. Unlike the issuance of the currency by the centralized authority, the Bitcoin is created by the mining process named Proof of work. After the successful creation of the block, the miner is rewarded with few Bitcoin. This incentive maintains the encouragement among the minors to join the network, contribute their CPU energy, and get Bitcoin as a reward. Transactions are hashed and encrypted using the secured hash algorithm or SHA-256 if all the minor validate the transactions.

The first genesis block of Bitcoin generated when Nakamoto sent ten Bitcoins to the noted programmer Finney and completed the first transaction. First time Florida bought $25 worth two pizzas in the exchange of 10k Bitcoin [49]. Now Bitcoin market cap is more than $210 billion [47]. Bitcoin is the digital currency generated in the distributed nodes. Bitcoin transactions are stored in public shared ledger in a distributed peer to peer network. The consensus process and the incentive mechanism makes it the most popular cryptocurrency among investors.

The minors can participate in the competition individually, or they can join the mining pool to get more probability to win. The investor can also buy the goods from a merchant with an exchange of Bitcoin. Both the seller and buyer can use the Bitcoin software to record the exchange of Bitcoin and goods. The transactions are initiated, and their hash value is calculated.

The transactions are broadcasted throughout the network. The block is created, and the transactions are combined into that valid block. The timestamp is used to verify the existence of transactions to prevent double-spending. Now the current block is hashed. The target is to find hash such that hash must have a fixed number of zeros as the prefix. To achieve this, the Nonce is used. This task is the difficulty to create the new hash of the block [65]. This mechanism consumes the energy but also verifies the block easily [66].

Bitcoin software can be downloaded on the computer and request a wallet. In a blockchain network, nodes can validate transactions by solving the complex algorithms, add them into their ledger, and broadcast the new block to the bitcoin network. After every 10 minutes, a block which contains the accepted transactions is created and added to the bitcoin blockchain network. This block is published to all nodes, without requiring any central oversight. Miners keep consistency in the network. The size of the bitcoin network block is 1 MB. This bitcoin protocol includes a higher level of security from the attacks like unauthorized spending, double spending, race attack, history modification, deanonymisation of clients. Bitcoin's implementation of public-private key cryptography removes unauthorized spending. For example; when A sends a bitcoin to B, B becomes the new owner of the bitcoin. C observing this transaction want to spend the bitcoin B just received, but C cannot sign the transaction without the knowledge of B's private key. Double spending is the case when a user sends the same currency to more than one user. This situation is solved by using a timestamp and tracking the transaction in the ledger in the blockchain network. Race attack is when two transactions are created with the same funds at the same time, intending to spend those funds twice. Memory and computational power are required for proof construction. This requirement of proof restricts the number of transactions to be validated in a given period. Each mined block produces new currency units, the total of which is finite. It is necessary to slow down the rate of production to prevent untimely exhaustion. In the case of Bitcoin, the number of newly produced bitcoins is halved every 210,000 blocks. Therefore, at most, there can be 21 million Bitcoins [1]. The success of Bitcoin has inspired the industry to create many other cryptocurrencies for a specific purpose or for multi-purpose.

B. Bitcoin Cash

Bitcoin Cash is created as a solution for scalability. It has a block size of 8 MB. The greater size of the block can have a higher number of transactions. Bitcoin permits only one block to be created in 10 minutes with a size of 1 MB. However, in Bitcoin Cash, the block is verified with a size of 8 MB, i.e., a higher number of transactions, and thus it boosts up the speed of transaction processing and saves the time to confirm the transactions. Due to the constant block size and fixed average confirmation time per block, Bitcoin has had very lower transaction throughput as less than ten transaction per second [50]. Bitcoin Cash and Ethereum both have higher throughput than Bitcoin, but still, their throughput is very less than the traditional payment systems like Paypal and Visa [50].

C. Zilliqa

Zilliqa has the mechanism to divide the task of mining to the sharded blockchain network. The blockchain network is divided into subnetworks called a shard. The mining process is performed on this shard in parallel. Thus the validation and verification can be done in parallel in the effectively lesser time. It can get a higher efficiency if there are a very high number of transaction to be mined [51].

D. ZCash

ZCash has higher security and privacy rather than Bitcoin. It has more powerful cryptographic techniques zero-knowledge proof [53], [54].

E. USDT

USDT is designed to maintain the stability between the cryptocurrency and the real currency of the country. This cryptocurrency is the best alternative to the real national currency.

F. Litecoin

There are only a few differences between Bitcoin and Litecoin like Litecoin uses the script and its average block creation time is 2.5 minute and allows the faster block confirmation. Litecoin is more complicated than Bitcoin.

G. Peercoin

Peercoin is based on proof of stake. Proof of stake includes additional proof of work in its consensus algorithm and uses the double SHA-256 algorithm. Peercoin is the first cryptocurrency that used PoS. The coinage is reflected as output, and therefore, the mining probability is higher for those nodes which keep the older coin.

H. Permacoin

Permacoin uses the consensus algorithm proof of retrievability. It requires clients to invest in large space, not just computational resources like Bitcoin. Permacoin system involves an alternative scratch-off puzzle based on Proofs-of retrievability (PORs). It requires large memory and therefore, the designer of Permacoin proposed storage instead of CPU cycles to secure the network for a new method to back up the data. Miners have to solve the mathematical problem, and this mathematical problem is called a scratch-off puzzle. It uses the considerable storage for decentralization of files.

I. Ethereum

Ethereum is based on EtHash. EtHash is the hashing algorithm developed by Ethereum developer. EtHash is ASIC resistant, and it has the property of memory hardness, or it is more follower of the fastness of data transferring in the memory. Ethereum uses proof of work. However, it does not use the previous hash algorithm. It mitigates the problem of mining centralization, in which a small mining operation can acquire a disproportionately large amount of power to manipulate the network. The current cryptocurrencies, according to their market cap, are listed in Table 3.

7. Blockchain Applications

Blockchain technology has most successful application in the finance sector as cryptocurrencies. However, there are many other sectors where blockchain can be utilised. There are many research projects ongoing to explore the adoption of blockchain for different applications. The critical characteristics of blockchain show the use of blockchain in the Internet of things, cloud and edge computing, big data, public and social services, electricity and in many more fields. The attracted feature of blockchain is distributed immutable information, and data is verified with the decentralized consensus mechanism. The broad spectrum of blockchain applications can be classified into three major sector- IoT, Big data, Cloud and Edge computing. The other sectors are economics and markets, business solution, smart contract and automation, traceability in supply chains, medical informatics, communications and networking, and others.

A. Internet of Things (IoT)

The smart devices using IoT have shown their potential in the different sectors. These devices are capable of working for a different purpose. Still, the system is not capable of handling huge data and the problematic complex computing mechanism such as described in [67-74]. The electronic devices such as sensors are increasing in the number, but their capability to collect and transmit the data is not efficient. The blockchain can be applied to the distributed devices to increase the security and privacy in IoT. Figure 10 shows the various sectors where the Blockchain technology can be applied.

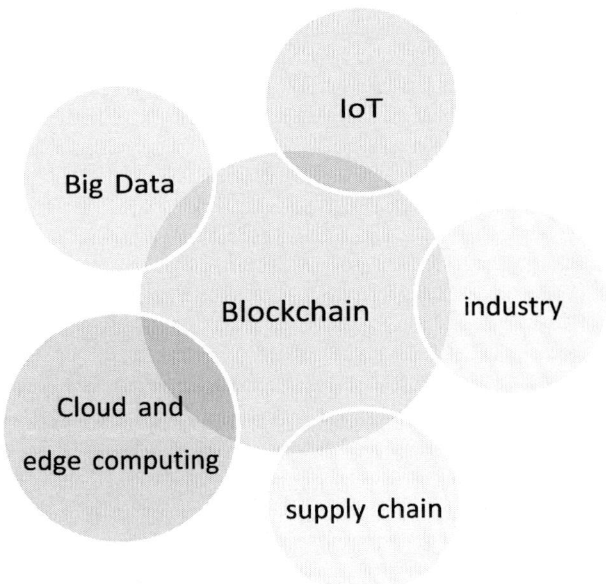

Figure 10. Different sectors of Blockchain.

Table 3. Cryptocurrency according to market cap [107]

List of Cryptocurrencies in Market					
S. No.	**Cryptocurrency**	**Abbreviation**	**Market Cap**	**Price (USD)**	**Year Introduced**
1	Bitcoin	BTC	$212.08 B	$11,871.16	2009
2	Ethereum	ETH	$24.14 B	$225.14	2015
3	XRP	XRP	$13.19 B	$0.307622	2013
4	Litecoin	LTC	$5.68 B	$90.10	2011
5	Bitcoin Cash	BCH	$5.99 B	$334.08	2017
6	EOS	EOS	$3.88 B	$4.19	2017
7	Binance coin	BNB	$4.39 B	$30.65	2017
8	Bitcoin SV	BSV	$2.58 B	$144.45	2018
9	Tether	USDT	$4.04 B	$1.00	2015
10	Cardano	ADA	$1.32 B	$0.050876	2019
11	Stellar	XLM	$1.51 B	$0.077167	2014
12	Tron	TRX	$1.47 B	$0.022080	2017
13	Monero	XMR	$1.62 B	$94.56	2014
14	UNUS SED LEO	LEO	$1.25 B	$1.25	2019
15	Dash	DASH	$949.28 M	$105.83	2014
16	Cosmos	ATOM	$657.13 M	$3.45	2017
17	IOTA	MIOTA	$757.59 M	$0.272561	2016
18	NEO	NEO	$768.34 M	$10.89	2014
19	Ethereum Classic	ETC	$688.90 M	$6.12	2015
20	NEM	XEM	$559.87 M	$0.062208	020
21	Tezos	XTZ	$901.02 M	$1.36	2017
22	Ontology	ONT	$485.88 M	$0.911167	2018
23	Zcash	ZEC	$451.59 M	$63.19	2016
24	Maker	MKR	$545.49 M	$545.49	2018
25	Chainlink	LINK	$808.43 M	$2.31	2019
26	Bitcoin Gold	BTG	$282.64 M	$16.14	2019
27	Crypto.com Chain	CRO	$409.45 M	$0.047343	2019
28	BasicAtt. coin	BAT	$277.42 M	$0.217290	2017
29	VeChain	VET	$268.67 M	$0.004845	2017
30	Dogecoin	DOGE	$352.57 M	$0.002921	2013
31	Qtum	QTUM	$277.35 M	$2.89	2017
32	USD Coin	USDC	$433.62 M	$1.00	2018
33	Decred	DCR	$290.92 M	$28.51	2016
34	OmisGO	OMG	$203.36 M	$1.45	2017
35	BitTorrent	BTT	$156.50 M	$0.000738	2019
36	Bitcoin Diamond	AL	$144.27 M	$0.773590	2017
37	Lisk	LSK	$151.33 M	$1.27	2016
38	Holo	HOT	$123.56 M	$0.000928	2018
39	Waves	WAVES	$135.06 M	$1.35	2016
40	Ravencoin	RVN	$162.17 M	$0.039069	2018
41	True USD	TUSD	$201.09 M	$1.00	2018

List of Cryptocurrencies in Market					
S. No.	**Cryptocurrency**	**Abbreviation**	**Market Cap**	**Price (USD)**	**Year Introduced**
42	Pundi X	NPXS	$125.73 M	$0.000536	2018
43	0x	ZRX	$117.12 M	$0.195178	2017
44	Nano	NANO	$143.72 M	$1.08	2017
45	Augur	REP	$120.97 M	$11.00	2015
46	Bytom	BTM	$106.96 M	$0.106690	2017
47	Bytecoin	BCN	$91.57 M	$0.000497	2014
48	Huobi Token Islands	HT	$253.53 M	$5.07	2018
49	RIF Token	RIF	$66.47 M	$0.14	2019
50	Elastos	ELA	$49.51 M	$3.13	2018
51	Horizen	ZEN	$47.91 M	$6.79	2017
52	Electroneum	ETN	$45.59 M	$0.004669	2017
53	SOLVE	SOLVE	$43.99 M	$0.134472	2019
54	Revain	R	$43.93 M	$0.090672	2017
55	Ecoreal Estate	ECOREAL	$43.61 M	$0.208019	2018
56	Stratis	STRAT	$44.68 M	$0.449327	2016
57	Decentraland	MANA	$43.44 M	$0.041368	2017
58	Zilliqa	ZIL	$76.20 M	$0.008771	2018
59	BitShares	BTS	$122.43 M	$0.044707	2014
60	Aurora	AOA	$120.00 M	$0.018341	2018
61	DigiByte	DGB	$119.03 M	$0.009818	2014
62	MonaCoin	MONA	$123.14 M	$1.87	2014
63	KuCoin Shares	KCS	$162.37 M	$1.83	2017
64	ICON	ICX	$108.51 M	$0.221242	2017
65	HyperCash	HC	$98.30 M	$2.26	2017
66	Komodo	KMD	$97.58 M	$0.845725	2017
67	Insight Chain	INB	$102.89 M	$0.294060	2018
68	Paxos Standard Token	PAX	$202.03 M	$1.00	2018
69	GXChain	GXC	$104.78 M	$1.61	2017
70	Verge	XVG	$83.42 M	$0.005257	2014
71	HedgeTrade	HEDG	$265.24 M	$0.919703	2019
72	IOST	IOST	$125.07 M	$0.010411	2018
73	Aeternity	AE	$78.85 M	$0.283804	2017
74	Siacoin	SC	$104.75 M	$0.002509	2015
75	Steem	STEEM	$68.14 M	$0.200765	2016
76	Ardor	ARDR	$63.59 M	$0.063659	2016
77	THETA	THETA	$108.54 M	$0.124691	2018
78	ABBC Coin	ABBC	$92.40 M	$0.166698	2018
79	Egretia	EGT	$178.72 M	$0.042369	2018
80	Mixin	XIN	$102.00 M	$223.95	2018
81	Nash Exchange	NEX	$108.25 M	$2.99	2019
82	Enjin Coin	ENJ	$52.63 M	$0.067815	2017
83	VestChain	VEST	$64.70 M	$0.009141	2018

Table 3. (Continued)

List of Cryptocurrencies in Market					
S. No.	**Cryptocurrency**	**Abbreviation**	**Market Cap**	**Price (USD)**	**Year Introduced**
84	aelf	ELF	$48.99 M	$0.098024	2017
85	Status	STF	$70.05 M	$0.020184	2017
86	Zcoin	XZC	$66.82 M	$8.25	2016
87	Crypto.com	MCO	$62.26 M	$3.94	2017
88	MaidSafeCoin	MAID	$80.60 M	$0.178101	2014
89	Golem	GNT	$54.48 M	$0.056493	2016
90	Energi	NRG	$136.63 M	$7.03	2018
91	WAX	WAX	$59.54 M	$0.063148	2017
92	Dai	DAI	$77.12 M	$1.00	2017
93	Grin	GRIN	$53.37 M	$3.03	2019
94	Project Pai	PAI	$41.57 M	$0.028565	2018
95	Maximine Coin	MXM	$52.90 M	$0.03	2018
96	NULS	NULS	$35.07	$0.475384	2017
97	Waltonchain	WTC	$64.42 M	$1.53	2017
98	ODEM	ODE	$51.34 M	$0.22	2018
99	Clipper Coin	CCCX	$52.15 M	$0.013	2018
100	NEXT	NET	$69.45 M	$1.38	2013

The blockchain technology can also be used to monitor the energy resources and consumption using the blockchain-based energy resource management application for cloud data centers. The authors [75], [97] has investigated the security and privacy issue in IoT and proposed the solution by creating the integrated framework of blockchain and IoT. The authors [22] came up with the possible features of the blockchain and IoT. There are many applications of blockchain in IoT. The authors [23], [24] focused on the application of blockchain in IoT. The research work [25] has categorized the applications and use cases of Blockchain for IoT.

B. Blockchain in Big Data

Initially, it was difficult to collect, store, and operate a large amount of data, but the advancement of machine learning and data mining techniques have paved the way for big data analysis [76],[77]. The rise of Big data with IoT still suffers from several issues of security, privacy, and centralized trust. The blockchain technology can be used to manage the distributed data and decentralization of the data processing. Unlike other techniques, blockchain provides the decentralized mechanism to store, manage, and process big data. It solves the privacy issues and empowers data security in Big data analysis [78]. The work [79] focused on how to anonymize dataset trading using blockchain technology. They designed a blockchain technology-based application using Hyperledger Fabric in which the nodes, being data brokers, collects and verifies the transactions in each data transfer. The blockchain-based platform achieves higher security in data distribution and data trading market. The authors [81] has negotiated the security issues in fog computing or centralized server-based cloud system. The fog computers are located such that only secured, and reliable data is sent to the centralized cloud systems. Still, drawbacks like the shutdown and reprocessing in the fog computers create significant issues to the cloud system. Blockchain technology brings out the security features to share and control the authenticated transactions by digital signature and consensus among fog computers. The research work [81] designed blockchain-based fog computing platform to solve the security issues of the centralized fog computing and cloud system. When a fog computing is shutdown, the distributed consensus process of blockchain successfully restores all the transactions to the fog computing.

Centralized database system suffers from security vulnerabilities like a single point of failure, IP spoofing, and Sybil attack [54]. The work [80] has proposed blockchain-based data management and searching method to ensure security issues. The digital signature to authenticate the information and check the user identity. The signature validation can prevent IP spoofing. The authors [80] developed such a system that can locate the IoT devices by having all the combined UUID that is resulted from gateway and IP address assigned IoT devices. The transaction includes the name with IP address and port number with the signature, and therefore the transactions can be traced by the name and authenticated by the signature.

C. Blockchain in Cloud Computing and Edge Computing

Edge computing, the processing of the data at the network edge, has shown its potential to improve the response time, battery life, bandwidth, data safety, and privacy. The services of the cloud is pushed into the edge network. The research work [82], [77] has listed the edge computing for smart home, smart city, and the data sharing and collaboration between networks of long-distance. However, it suffers from the challenges of system integration, resource management, the programmability of edge computing, naming mechanism, security and privacy issues in transmission, storage and computation, etc. The authors [82-85] proposed the cloud computing-based system for enhancing the cybersecurity for large enterprise networks with reduced operational delays and can detect threats by parallel cloud computation for both the signature and anomaly-based detection. Cloud and edge computing both have a significant application, but with blockchain the distribution mechanism and the consensus process, the challenges of cloud and edge computing are resolved. The service contract management of blockchain for cloud and edge computing allows the programmability for the users. Blockchain for the distributed centralized large datacenters brought the consensus and decentralized layer.

The authors [86] proposed the cloud architecture based on blockchain technology with fog computing and software-defined networking for the efficient management of the data produced by the cloud and edge computing. This proposed blockchain-based architecture provides high security. Scalability and resiliency with the very low latency between the computing resources and IoT devices. The cloud computing cost and the number of trusted third parties can be reduced significantly by this architecture.

The authors in [87] have shown the data unavailability, higher operational cost, and data security issues in the traditional cloud storage and brought a blockchain-based model with keyword search services to solve all these issues. The encrypted client data is distributed to the cloud nodes to ensure the data availability using cryptographic algorithms with the private keyword search feature for the owner. They built a system to enable the outsourcing of data storage to a distributed network by using the proof of retrievability consenses. Blockchain is used to process and prove the anonymous credential with the private keyword search feature. The work [88] has enlighted the edge devices less efficient for computational resources and the available bandwidth leading to fog or cloud. By measuring the network latency, they have shown that the use of fog and cloud to make the IoT application with the blockchain technology.

D. Blockchain in Supply Chain and Logistics

The blockchain technology can distribute information throughout the whole supply chain network with product tracking, traceability facility, and improved quality, sustainability, and flexibility. The blockchain-based supply chain platform can improve cost, time, and risk management. The authors in [89] have presented the use of blockchain for increased transparency and accountability for different phases of product development. The blockchain technology brings out the powerful features to trace the product with transparent, secure, and accurate information. The smart contract can be used to automate the payment when a product is returned to the issuer or seller. The supply chain can have real-time tracking verified by consensus process and connect all the members on the same platform.

In [91], the authors designed a blockchain-based system using proof of concept consensus to prove the ownership of products along with radio frequency identification or RFID to diminish the counterfeits in the supply chain. The authors in [92] proposed a blockchain-based agri-food supply chain traceability system using RFID. In work, they proved the traceability and trust in the entire agri-food supply chain by transferring, processing and distributing the authentic data [93]. The digital records created with the blockchain, RFID, and IoT can deduce the conflicts and solve the judgment in the supply chain. All the critical information about materials, transaction processes, product quality, location, and quality is maintained on the blockchain-based supply chain process.

The authors [89] have described the problems in product deletion management in the current supply chain operations like origination, operation, manufacturing, logistics, and reverse logistics. They built blockchain-based product management to empower product deletion and rationalization.

E. Industry Perspective

The traditional industry uses the intermediaries for registering the transactions, confirming and removing the duplicate transactions. The blockchain-based platform can perform these task of registering, confirming, and storing the transactions with higher efficiency in minimum time. Blockchain can be used for the insurance policies, verifying the claims via smart contract to reduce the overall cost.

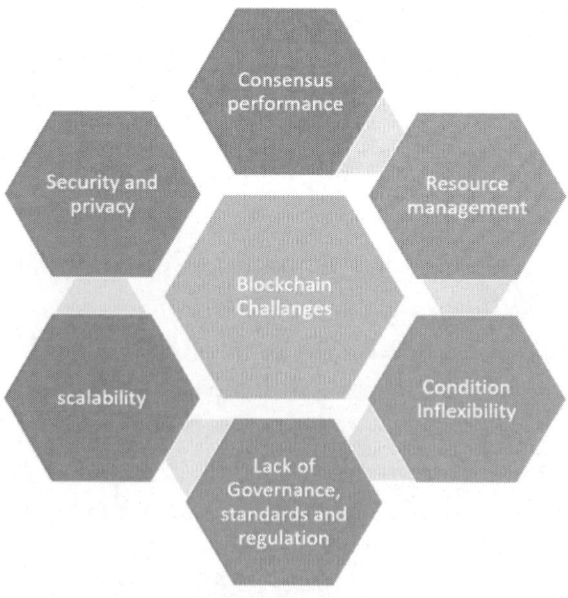

Figure 11. Challenges of Blockchain technology.

8. CHALLENGES OF BLOCKCHAIN TECHNOLOGY

Blockchain technology can be used in a broad spectrum of applications to secure the data with increased reliability without requiring any central trusted authority. It has attracted the researchers due to its key characteristics decentralization, immutability, consistency, and security. Though, it still suffers from a few challenges like scalability, storage management, consensus performance and resources management, etc. Figure 11 shows the significant challenges of Blockchain technology.

A. Scalability

The prefixed block size and block creation time are efficient for a fixed number of transactions processing, but a high number of transactions can cause slower transaction processing. Several blockchain applications are suffering from the scalability issue such as Bitcoin block size is 1MB and average block confirmation time is 10 minutes, whereas Ethereum has the average block confirmation time 16 seconds [94-95]. For the high number of transaction processing, the block confirmation time must be low, but to have the security from being attacked by the attacker, this average time should be high. To solve this problem, there are a few solutions proposed by the researchers. The authors [7] proposed the solutions like on-chain to change on main, off the chain for change on the main chain after transaction processing. Also, side chain to change the assets of the different side chain, child chain to record the result in parent chain and interchain for the communication between the chains. In [97], the authors have proposed solutions like lighting protocols, sharding, etc. This work [100] has analyzed the better performance of Ethereum by managing the scalability issues.

B. Security and Privacy Issues

There is an urgent need to research for tor offers and beyond tor such that the privacy issues can be tackled in blockchain [97]. To create and manage the digital identities, the rules and policies are to be restricted to keep the privacy for controls and ownership while accepting the blockchain model.

C. Storage Management

Blockchain ledger contains all the transaction starting from geniuses block to the latest mined blocks with all the side and interchains. The blockchain provides higher security by storing all transactions in a ledger. This ledger is distributed to each node of the blockchain network. This redundancy spends nonpredictable storage. The storage management depends on the activity of the nodes of the network.

D. Consensus Performance

There are several consensus algorithms proposed for different applications. The consensus for a specific purpose can create scalability, efficiency, and privacy issues. The efficiency of the consensus process decreases when the number of nodes in the network increases.

E. Resource Management

The resources used to mine the block are less in the beginning as there are very fewer transactions, but for the vast number of transactions, the spent resources are very high. If miners spend the resources, they must get a satisfactory reward. Therefore it becomes a challenge to maintain compatibility between the miners and the spent resources. Resources consumption should be managed effectively.

F. Conditions Inflexibility

The smart contract is designed in a specific programming script languages that do not provide comfortable functions to the programmer. The programmers have shown the indeed requirement of the function to get the flexibility to create the programs so that the contract evaluation may be easier. While creating a smart contract, the complex understanding and compilation process may create unpredictable worst results. Inflexibility restricts to test the contract against the bugs and prevent the attacks. The authors [101] has shown the hacking event of the smart contract built on Ethereum [102-103], and the work [104] have shown the vulnerabilities and the possible improvement while creating the smart contract.

G. Lack of Governance, Standards, and Regulations

It is indeed a requirement to standardize the blockchain for its integration, interoperability, governance, and sustainability, etc. The blockchain development must follow the rules, laws, policies, and regulations of the government. It is challenging to manage the governance of the blockchain platform among different participants. The authors [105] has raised the questions on the blockchain standards and standardization importance to maintain sustainability and trust. This work [106] has reviewed the blockchain regulation and has measured the performance factor.

CONCLUSION

This chapter has presented the underlying architecture, mechanism, applications, challenges, and opportunities of the blockchain technology. The chapter has described the key characteristics of blockchain, such as decentralized, immutable, distributed, and secured. The comparison between different versions and variants of the blockchain technology as well as the fundamental difference between the various consensus algorithms has been described in details in this chapter. The basic principle of cryptocurrencies and their applications are presented for naive readers. Moreover, the chapter has depicted the current research and industrial challenges to adopt the blockchain for different application in terms of scalability, privacy and security, storage and resource management, and lack of governance and standardization.

REFERENCES

[1] Nakamoto, Satoshi. "*Bitcoin: A peer-to-peer electronic cash system.*" (2008).

[2] Dinh, TTA; Liu, R; Zhang, M; Chen, G; Ooi, BC; Wang, J. "Untangling blockchain: A data processing view of blockchain systems," *IEEE Transactions on Knowledge and Data Engineering*, vol. 30, no. 7, pp. 1366-1385, 2018.

[3] Tschorsch, F; Scheuermann, B. "Bitcoin and beyond: A technical survey on decentralized digital currencies," *IEEE Communications Surveys Tutorials*, vol. 18, no. 3, pp. 2084-2123, 2016.

[4] Kosba, A; Miller, A; Shi, E; Wen, Z; Papamanthou, C. "Hawk: The blockchain model of cryptography and privacy-preserving smart contracts," in *IEEE Symposium on Security and Privacy (SP)*, San Jose, CA, May 2016, pp. 839-858.

[5] Bonneau, J; Miller, A; Clark, J; Narayanan, A; Kroll, JA; Felten, EW. "Sok: Research perspectives and challenges for bitcoin and cryptocurrencies", in *IEEE Symposium on Security and Privacy, San Jose, CA*, May 2015, pp. 104-121.

[6] Yeow, K; Gani, A; Ahmad, RW; Rodrigues, JJPC; Ko, K. "Decentralized consensus for edge-centric internet of things: A review, taxonomy, and research issues", *IEEE Access*, vol. 6, pp. 1513-1524, 2018.

[7] Kim, Soohyeong, Yongseok Kwon, Sunghyun Cho. "A survey of scalability solutions on blockchain." *International Conference on Information and Communication Technology Convergence (ICTC). IEEE, 2018.*

[8] Conoscenti, M; Vetro, A; De Martin, JC. "Blockchain for the Internet of Things: a systematic literature review." in *IEEE/ACS 13th International Conference of Computer Systems and Applications (AICCSA)*, pages 1-6, Nov 2016.

[9] Polyzos, GC; Fotiou, N. "Blockchain-assisted information distribution for the Internet of Things." In *IEEE International Conference on Information Reuse and Integration (IRI)*, pages 75-78, August 2017.

[10] "*Proof of authority chains*," Jan. 2018. [Online]. Available: https://github.com/paritytech/parity.

[11] Gueron, S; Johnson, S; Walker, J. "SHA-512/256," *Eighth International Conference on Information Technology: New Generations*, Las Vegas, NV, 2011, pp. 354-358.

[12] Lamba, S; Sharma, M. "An Efficient Elliptic Curve Digital Signature Algorithm (ECDSA)," *International Conference on Machine Intelligence and Research Advancement*, Katra, 2013, pp. 179-183.

[13] Chen, L; Xu, L; Gao, Z; Lu, Y; Shi, W. "Protecting Early Stage Proof-of-Work Based Public Blockchain," *48th Annual IEEE/IFIP International Conference on Dependable Systems and Networks Workshops (DSN-W)*, Luxembourg City, 2018, pp. 122-127.

[14] Alkandari, AA; Al-Shaikhli, IF; Alahmad, MA. "Cryptographic Hash Function: A High Level View, "*International Conference on Informatics and Creative Multimedia, Kuala Lumpur*, 2013, pp. 128-134.

[15] Kaur, R; Kaur, A. "Digital Signature," *International Conference on Computing Sciences, Phagwara*, 2012, pp. 295-301.

[16] Dziembowski, S; Faust, S; Kolmogorov, V; Pietrzak, K. "Proofs of space," in *Advances in Cryptology & CRYPTO: 35th Annual*

Cryptology Conference, Santa Barbara, CA, Aug. 2015, pp. 585-605.

[17] Chowdhury, MJM; Colman, A; Kabir, MA; Han, J; Sarda, P. "Blockchain Versus Database: A Critical Analysis," *17th IEEE International Conference On Trust, Security And Privacy In Computing And Communications/12th IEEE International Conference On Big Data Science And Engineering (TrustCom/BigDataSE)*, New York, NY, 2018, pp. 1348-1353.

[18] Dorri, A; Kanhere, SS; Jurdak, R; Gauravaram, P. "Blockchain for IoT security and privacy: The case study of a smart home," in *IEEE International Conference on Pervasive Computing and Communications Workshops (PerCom Workshops)*, March 2017, pp. 618-623.

[19] Zhang, Y; Wen, J. "The IoT electric business model: Using blockchain technology for the internet of things," *Peer-to-Peer Networking and Applications*, vol. 10, no. 4, pp. 983-994, Jul 2017.

[20] Conoscenti, M; Vetro, A; De Martin, JC. "Blockchain for the internet of things: A systematic literature review," in *IEEE/ACS 13th International Conference of Computer Systems and Applications (AICCSA)*, Nov 2016, pp. 1-6.

[21] Banerjee, M; Lee, J; Choo, KKR. "A blockchain future for internet-of-things security: a position paper," *Digital Communications and Networks*, vol. 4, no. 3, pp. 149-160, 2018.

[22] Reyna, A; Martn, C; Chen, J; Soler, E; Daz, M. "On blockchain and its integration with IoT. Challenges and opportunities," *Future Generation Computer Systems*, vol. 88, pp. 173-190, 2018.

[23] Fernandez-Carames, TM; Fraga-Lamas, P. "A review on the use of blockchain for the internet of things," *IEEE Access*, vol. 6, pp. 32 979-33 001, 2018.

[24] Ali, MS; Vecchio, M; Pincheira, M; Dolui, K; Antonelli, F; Rehmani, MH. "Applications of blockchains in the internet of things: A comprehensive survey," *IEEE Communications Surveys Tutorials*, pp. 1-42, 2018 (early access). [Online]. Available: https://doi.org/10.1109/COMST.2018.2886932.

[25] Panarello, A; Tapas, N; Merlino, G; Longo, F; Puliafito, A. "Blockchain and iot integration: A systematic survey," *Sensors*, vol. 18, no. 8, 2018. [Online]. Available: http://www.mdpi.com/ 1424-8220/18/8/2575.

[26] Wood, G. "Ethereum: A secure decentralised generalised transaction ledger", *Ethereum Project Yellow Paper*, vol. 151, pp. 1-32, Apr. 2014.

[27] Cox, T. "*Eos. io technical white paper*" in GitHub repository, 2017.

[28] Mazieres, D. "*The stellar consensus protocol: A federated model for internet-level consensus*", Stellar Developer Foundation.

[29] Mogan, J. Quorum. *Advancing Blockchain Technology*, 2018, [online] Available: https://www.jpmorgan.com/country/ US/EN/ Quorum.

[30] Schwartz, D; et al., *The ripple protocol consensus algorithm*, San Francisco, CA, USA, 2014.

[31] Cachin, C. "Architecture of the hyperledger blockchain fabric", *Proceedings Workshop Distributed Cryptocurrencies Consensus Ledgers*, pp. 310, 2016.

[32] Mainelli, M; von Gunten, C. "*Chain Of A Lifetime: How Blockchain Technology Might Transform Personal Insurance*." Long Finance, London, U.K. [Online]. Available: http://archive. longfinance.net/images/Chain_Of_A_Lifetime_December2014.pdf

[33] Cohn, A; West, T; Parker, C. "Smart after all: Blockchain, smart contracts, parametric insurance, and smart energy grids," *Georgetown Law Technology Review*, vol. 1, no. 2, pp. 273-304, 2017.

[34] Kwon, Jae. "*Tendermint: Consensus without mining*." Draft v. 0.6, fall (2014).

[35] *Ethereum Project. Accessed: Jan. 2019.* [Online]. Available: https://www. ethereum.org/.

[36] Hyperledger. Accessed: July. 2019. [Online]. Available: https://www. hyperledger.org/.

[37] Hijro. Accessed: Nov. 2018. [Online]. Available: https://hijro.com/.

[38] Tierion. Accessed: Feb. 2019. [Online]. Available: https://tierion.com/.

[39] Prisco, G. (Apr. 2016). The Blockchain for Healthcare: Gem Launches Gem Health Network With Philips Blockchain Lab. *BitCoin Magazine*. Accessed: Feb. 2019. [Online]. Available: https://bitcoinmagazine.com/articles/the-blockchain-for-heathcare-gemlaunches-gem-health-network-with-philips-blockchain-lab-1461674938/.

[40] Steiner, J; Baker, J. "Blockchain: The solution for transparency in product supply chains." *Project Provenance Ltd., London, U.K., White Paper,* 2015. [Online]. Available: https:// www.provenance.org/whitepaper.

[41] Szabo, N. "Smart contracts: Building blocks for digital markets," *in Extropy*, 16, 1996.

[42] Peters, Daniel, et al. "Blockchain applications for legal metrology." *IEEE International Instrumentation and Measurement Technology Conference (I2MTC)*. 2018.

[43] Cai, Wei, et al. "Decentralized applications: The blockchain-empowered software system." *IEEE Access*, 6 (2018), 53019-53033.

[44] Michel Goossens, Frank Mittelbach, Alexander Samarin. *The L^AT_EX Companion*. Addison-Wesley, Reading, Massachusetts, 1993.

[45] Yuan, Yong, Fei-Yue Wang. "Blockchain and cryptocurrencies: Model, techniques, and applications." *IEEE Transactions on Systems, Man, and Cybernetics: Systems,* 48.9 (2018), 1421-1428.

[46] Modiri, N. "The ISO reference model entities," *IEEE Network Magzine*, vol. 5, no. 4, pp. 24-33, Jul. 1991.

[47] *Cryptocurrency Monitoring Website*. Accessed: Nov. 24, 2015. [Online]. Available: http://coinmarketcap.com/.

[48] Farell, Ryan. "*An analysis of the cryptocurrency industry*." (2015).

[49] *The Street Report*. Accessed: July. 09, 2019. [Online]. Available: https://www.thestreet.com/investing/bitcoin/bitcoinhistory-14686578.

[50] Chow, Sherman SM; et al. "Sharding Blockchain." *IEEE International Conference on Internet of Things (iThings) and IEEE*

Green Computing and Communications (GreenCom) and IEEE Cyber, Physical and Social Computing (CPSCom) and IEEE Smart Data (SmartData) IEEE, 2018.

[51] (2016). Zilliqa. [Online]. Available: https://zilliqa.com/.

[52] Chomsiri, Thawatchai, Kan Kongsup. "P Coin: High Speed Cryptocurrency Based on Random-Checkers Proof of Stake." *Joint* IEEE *10th International Conference on Soft Computing and Intelligent Systems (SCIS) and 19th International Symposium on Advanced Intelligent Systems (ISIS)*, 2018.

[53] Kshetri, Nir. "Cryptocurrencies: Transparency Versus Privacy [Cybertrust]." *Computer,* 51.11 (2018), 99-111.

[54] Douceur, JR. "The sybil attack," in *First International Workshop on Peer-to-Peer Systems, Cambridge, MA, Mar. 2002,* pp. 251-260.

[55] www.coinmarketcap.com, https://coinmarketcap.com/coins/views/all/, accessed: 2019-07-15.

[56] Miller, A; Juels, A; Shi, E; Parno, B; Katz, J. "Permacoin: Repurposing bitcoin work for data preservation," in *IEEE Symposium on Security and Privacy*, San Jose, CA, May 2014, pp. 475-490.

[57] Kopp, H; Bosch, C; Kargl, F. "Koppercoin â a distributed file storage with financial incentives," in *12th International Conference on Information Security Practice and Experience*, Zhangjiajie, China, Nov. 2016, pp. 79-93.

[58] Protocol Labs, "Filecoin: A decentralized storage network," *Protocol Labs, Technical Report*, Aug. 2017.

[59] Hopwood, D; Bowe, S; Hornby, T; Wilcox, N. "Zcash protocol specification," *Zerocoin Electric Coin Company, Technical Report*, Dec. 2017.

[60] Wood, G. "Ethereum: A secure decentralised generalised transaction ledger (eip-150 revision)," *Ethereum Project Yellow Paper*, vol. 151, 2017.

[61] Biryukov, A; Khovratovich, D. "Equihash: Asymmetric proof-of work based on the generalized birthday problem," *Ledger Journal*, vol. 2, pp. 1-30, Apr. 2017.

[62] Wang, Wenbo; et al. "*A survey on consensus mechanisms and mining management in blockchain networks*." arXiv preprint arXiv:1805.02707 (2018), pp. 1-33.

[63] Maung Maung Thin, WY; Dong, N; Bai, G; Dong, JS. "Formal Analysis of a Proof-of-Stake Blockchain," *23rd International Conference on Engineering of Complex Computer Systems (ICECCS), Melbourne, VIC*, 2018, pp. 197-200.

[64] Castro, M; Liskov, B. "Practical byzantine fault tolerance and proactive recovery," *ACM Transactions on Computer Systems*, vol. 20, no. 4, pp. 398-461, Nov. 2002.

[65] Gobel, J; Keeler, HP; Krzesinski, AE; Taylor, PG. Bitcoin blockchain dynamics: The selfish-mine strategy in the presence of propagation delay. *Performance Evaluation*, vol. 104, pp. 23 - 41, 2016.

[66] Minhaj Ahmad Khan, Khaled Salah. IoT security: Review, blockchain solutions, and open challenges. *Future Generation Computer Systems*, vol. 82, pp. 95 - 411, 2018.

[67] Lin, J; Yu, W; Yang, X; Yang, Q; Fu, X; Zhao, W. "A realtime en-route route guidance decision scheme for transportation-based cyber physical systems." *IEEE Transactions on Vehicular Technology*, 66(3), pp. 2551-2566, March 2017.

[68] Xu, G; Yu, W; Griffith, D; Golmie, N; Moulema, P. Toward integrating distributed energy resources and storage devices in smart grid. *IEEE Internet of Things Journal*, 4(1), 192-204, Feb 2017.

[69] Ming Zhao, Arun Kumar, Peter Han Joo Chong, Rongxing Lu, "A Comprehensive Study of RPL and P2P-RPL Routing Protocols: Implementation, Challenges and Opportunities", in *Peer-to-Peer Networking and Applications*, pp. 1-25, July 2016.

[70] Arun Kumar, Hnin Shwe, Kai-Juan Wong, Peter H. J. Chong, "Location-based Routing Protocols for Wireless Sensor Networks: a survey" *Wireless Sensor Network*, Vol. 9, pp. 25-72, 2017.

[71] Ming Zhao, Arun Kumar, Tapani Ristaniemi, Peter Han Joo Chong, "Machine-to-Machine Communication and Research Challenges: A

Survey", in *Wireless Personal Communications, Springer*, pp. 1-17, 2017.

[72] Arun Kumar, Ming Zhao, Kai-Juan Wong, Liang Yong Guan, Peter Han Joo Chong, "A Comprehensive Study of IoT and WSN MAC Protocols: Research Issues, Challenges and Opportunities", *IEEE Access*, Vol. 6, pp. 76228 – 76262, November 2018.

[73] Minglong Zhang, Nawaz Ali, Peter Han Joo Chong, Boon-Chong Seet, Arun Kumar, "A Novel Hybrid MAC Protocol for Safety Message Broadcasting in Vehicular Networks", *IEEE Transactions on Intelligent Transportation Systems*, September 2019.

[74] Dorri, A; Kanhere, SS; Jurdak, R; Gauravaram, P. "Blockchain for IoT security and privacy: The case study of a smart home." In *IEEE International Conference on Pervasive Computing and Communications Workshops (PerCom Workshops)*, pp. 618-623, March 2017.

[75] Xu, C; Wang, K; Guo, M. "Intelligent resource management in blockchain-based cloud datacenters." *IEEE Cloud Computing*, 4(6), 50-59, November 2017.

[76] Liang, F; Yu, W; An, D; Yang, Q; Fu, X; Zhao, W. "A survey on big data market: Pricing, trading and protection." *IEEE Access*, 6, 15132-15154, 2018.

[77] Hatcher, WG; Yu, W. "A survey of deep learning: Platforms, applications and emerging research trends." *IEEE Access*, vol. 6, pages 24411-24432, 2018.

[78] Karafiloski, E; Mishev, A. "Blockchain solutions for big data challenges: A literature review." In *17th International Conference on Smart Technologies (IEEE EUROCON'17)*, pages 763-768, July 2017.

[79] Kiyomoto, S; Rahman, MS; Basu, A. "On blockchain-based anonymized dataset distribution platform." In *IEEE 15th International Conference on Software Engineering Research, Management and Applications (SERA)*, pages 85-92, June 2017.

[80] Jung, MY; Jang, JW. "Data management and searching system and method to provide increased security for IoT platform." In

International Conference on Information and Communication Technology Convergence (ICTC), pages 873-878, Oct 2017.

[81] Jeong, Jun Woo, Bo Youn Kim, Ju Wook Jang, "Security and Device Control Method for Fog Computer using Blockchain." *ACM Proceedings of the International Conference on Information Science and System*, 2018.

[82] Shi, W; Cao, J; Zhang, Q; Li, Y; Xu, L. "Edge computing: Vision and challenges." *IEEE Internet of Things Journal*, 3(5), 637-646, Oct 2016.

[83] Yu, W; Liang, F; He, X; Hatcher, WG; Lu, C; Lin, J; Yang, X. A survey on the edge computing for the Internet of Things. *IEEE Access*, 6:6900-6919, 2018.

[84] Yu, W; Guobin Xu, Zhijiang Chen, Moulema, P. "A cloud computing based architecture for cyber security situation awareness." In *IEEE Conference on Communications and Network Security (CNS)*, pages 488-492, October 2013.

[85] Ming Zhao, Arun Kumar, G. G. Md. Nawaz Ali, Peter H. J. Chong, "A Cloud-based Network Architecture for Big Data Services", in Proceedings of *the 2nd IEEE International Conference on Big Data Intelligence and Computing (DataCom'16)*, Auckland, New Zealand, August 8-12, 2016.

[86] Sharma, PK; Chen, MY; Park, JH. "A software defined fog node based distributed blockchain cloud architecture for IoT." *IEEE Access*, vol. 6, 115-124, 2018.

[87] Do, HG; Ng, WK. "Blockchain-based system for secure data storage with private keyword search." In *IEEE World Congress on Services (SERVICES)*, pages 90-93, June 2017.

[88] Samaniego, M; Deters, R. "Blockchain as a service for IoT", In *IEEE International Conference on Internet of Things (iThings) and IEEE Green Computing and Communications (GreenCom) and IEEE Cyber, Physical and Social Computing (CPSCom) and IEEE Smart Data (SmartData)*, pages 433-436, December 2016.

[89] Zhu, Qingyun, Mahtab Kouhizadeh. "Blockchain Technology, Supply Chain Information, and Strategic Product Deletion

Management." *IEEE Engineering Management Review 47.1*: pp. 36-44, 2019.

[90] Kshetri, Nir. "Blockchains roles in meeting key supply chain management objectives." *International Journal of Information Management*, 39, pp. 80-89, 2018.

[91] Toyoda, Kentaroh; et al. "A novel blockchain-based product ownership management system (POMS) for anti-counterfeits in the post supply chain." *IEEE Access*, 5 (2017), 17465-17477.

[92] Tian, Feng. "An agri-food supply chain traceability system for China based on RFID blockchain technology." *13th international conference on service systems and service management (ICSSSM). IEEE*, 2016.

[93] Alekha Kumar Mishra, Arun Kumar, Asis Kumar Tripathy, Tapan Kumar Das, "A Two-Tailed Chain Topology in Wireless Sensor Networks for Efficient Monitoring of Food Grain Storage", in *Recent Findings in Intelligent Computing Techniques*, Vol 708, pp. 97-105, November 2018, Springer, Singapore.

[94] *Average block time of the Ethereum Network* - etherchain.org https://www.etherchain.org/charts/blockTime.

[95] *Bitcoin Block Time chart* https://bitinfocharts.com/comparison/bitcoin-confirmationtime.html.

[96] Chauhan, Anamika; et al. "Blockchain and scalability." *IEEE International Conference on Software Quality, Reliability and Security Companion (QRS-C).* IEEE, 2018.

[97] Henry, R; Herzberg, A; Kate, A. "Blockchain Access Privacy: Challenges and Directions," in *IEEE Security Privacy*, vol. 16, no. 4, pp. 38-45, July/August 2018.

[98] Karame, G; Capkun, S. "Blockchain Security and Privacy," in *IEEE Security Privacy*, vol. 16, no. 4, pp. 11-12, July/August 2018.

[99] Yu, Y; Li, Y; Tian, J; Liu, J. "Blockchain-Based Solutions to Security and Privacy Issues in the Internet of Things," in *IEEE Wireless Communications*, vol. 25, no. 6, pp. 12-18, December 2018.

[100] Rouhani, S; Deters, R. "Performance analysis of ethereum transactions in private blockchain," *8th IEEE International Conference on Software Engineering and Service Science (ICSESS)*, Beijing, 2017, pp. 70-74.

[101] Destefanis, G; Marchesi, M; Ortu, M; Tonelli, R; Bracciali, A; Hierons, R. "Smart contracts vulnerabilities: A call for blockchain software engineering" in *Proceedings International Workshop Blockchain Oriented Software Engineering (IWBOSE)*, March 2018, pp. 19-25.

[102] Luu, L; Chu, DH; Olickel, H; Saxena, P; Hobor, A. "Making smart contracts smarter," in *Proc. ACM SIGSAC Conf. Computer Communication Security*, 2016, pp. 254-269.

[103] Tsankov, P; Dan, A; Cohen, DD; Gervais, A; Buenzli, F; Vechev, M. (2018). "*Securify: Practical security analysis of smart contracts*."

[104] Tikhomirov, S; Voskresenskaya, E; Ivanitskiy, I; Takhaviev, R; Marchenko, E; Alexandrov, Y. "Smartcheck: Static analysis of ethereum smart contracts," in *Proceedings of IEEE/ACM 1st International Workshop Emerging Trends Software Engineering Blockchain (WETSEB)*, May/Jun. 2018, pp. 9-16.

[105] Anjum, A; Sporny, M; Sill, A. "Blockchain standards for compliance and trust," *IEEE Cloud Comput.*, vol. 4, no. 4, pp. 84-90, Jul./Aug. 2017.

[106] Kakavand, H; De Sevres, NK; Chilton, B. (Jan. 2017). *The Blockchain Revolution: An Analysis of Regulation and Technology Related to Distributed Ledger Technologies*. [Online]. Available: https://ssrn.com/abstract=2849251.

[107] Online [Available] https://coinmarketcap.com/coins/views/all/. Accessed on 08-08-2019; 18:10.

In: Vehicular Ad-Hoc Networks ISBN: 978-1-53618-038-1
Editors: G. G. Md. Nawaz Ali et al.

Chapter 6

OPTIMAL HOLE COVERAGE IN THE PRESENCE OF DUMB NODE IN WIRELESS SENSOR NETWORKS

Pushpendu Kar*

School of Computer Science, University of Nottingham,
Ningbo, Zhejiang, China

ABSTRACT

Wireless Sensor Networks (WSNs) are energy and resource constraint networks. So, activation of an optimum number of sensor nodes in a WSN is always desirable to increase the lifetime of the network. Sometimes, due to adverse environmental conditions, sensor nodes may exhibit dumb behavior. A dumb sensor node can continue its sensing operation but cannot communicate due to impairments from the adverse environmental conditions. The sensor nodes behave normally after the resumption of a favorable environmental condition. The dumb behavior of the sensor nodes results in a communication hole(s) in the network, which creates, increases, or decreases dynamically. The creation of hole(s) may lead to data loss and degrade network performance. In

* Corresponding Author's Email: pushpendu.kar@nottingham.edu.cn.

such a scenario, to resume network performance, we proposed *Dynamic Hole Coverage for Dumb Node (DHCDN)* scheme to mitigate the effects of dumb nodes from the WSN by activating the optimum number of sleeping sensor nodes in the network.

Keywords: wireless sensor network, dumb node, communication hole, coverage

Introduction

In the current advancement of MEMS, Wireless Sensor Network (WSN) performs a very interesting and important task in many fields. Wireless Sensor Network consists of combinations of a set of low power sensors [1]. This Wireless Sensor Network plays an important role in many fields like knowing the characteristics of an object, the position of an object, the behavior of an object, and tracking object. Some nodes have short-range of transmission, so intermediate nodes are acting as a relay node to forward sensed information to the sink [2]. Sensor nodes work collaboratively to measure their surroundings (i.e., light, pressure, temperature, humidity, vibration).

Sensors can be deployed in two ways: grid-based and scattered or random deployment [3]. Grid-based deployment is a pre-planned basis and random deployment has no pre-organized plan. In WSN, nodes participate to form a network in a collaborative manner i.e., active participation and collaboration are very important information of a good sensor network.

A sensor node can have four states: transmit, receive, idle and sleep. Among all these states, the sleep state consumes much less energy as compared to other states [4]. As the battery power of the sensor node is very limited, it should be the aim to keep the maximum number of nodes in sleep mode in the network. So, it is desirable to activate sensor nodes optimally to cover the total region i.e., all the deployed sensor nodes don't remain active during the whole time interval. Some nodes are in the sleep state, some are inactive state. Activation of nodes needs to be in such a way that they cover maximum area and create a minimum hole(s) in the network, which in turn, maximize the network lifetime. All these criteria fulfilled by reducing the overlapping sensing area of the sensor nodes. Among activated nodes, one or many nodes may arise the dumb behavior i.e., node can sense but cannot communicate with any other node due to some environmental reason, some external effects or technical disorder of the sensor nodes. If a node is dumb, it is useless to sense the area by the sensor node as it cannot send sensed information to other nodes. If the communication range of nodes does not cover some portion of the area, this region is called a communication hole. The communication area of the dumb node can be treated as a hole, as it senses the area but cannot communicate the sensed information with any other node.

RELATED WORK

In WSN, cooperation among nodes is a very important factor for the proper functioning and performance of the networks. The appearance of a hole(s) in a network loses cooperation among nodes and becomes major hinder in proper functioning and better performance of the network. In [5] hole creation and coverage by mobile sensor nodes with some cascading movement has discussed by Liu et al. Here, proxy-based bidding protocol is developed, which moves mobile sensors physically from small holes to the large hole and the destination is confirmed before that mobile sensor node moves virtually. This approach reduces the movement distance by half as well as reduces energy consumption. The Best Fit Node Policy (BFNP) has proposed in [6], which covers the hole area by activating closest inactive nodes. The activated nodes cover the center of the minimal coverage circle that circumscribed the hole polygon. In [7], Hole Avoiding In Advance Routing protocol (HAIR) has proposed to re-route data packets in advance to avoid the hole instead of bypassing the hole. Qing Fang et al. [8] proposed a technique for getting out of a stuck position by a packet. If there is a hole region in the network, the boundary nodes cannot forward the packet further. After reaching a packet at the node on the boundary of the hole, which gets stuck due to the hole region i.e., it can't be forward further, and this node is called a stuck node. There is some mechanism for finding other stuck nodes on the boundary of the same hole to forward the packet. Here, it is emphasized by defining and discovering holes in a sensor network. Honghai Zhang et al. [9] have discussed the minimum activation of nodes, which maintain coverage and connectivity. They have introduced a new technique Optimal Geographical Density Control (OGDC), which can maintain coverage without relating sensing and communication range. In the paper [10], the hole boundary nodes finding technique in a WSN has been discussed. The authors proposed an algorithm named Distributed Boundary Recognition Algorithm (DBRA) for this. According to the authors, the proposed algorithm will take less time and communication overhead. Wan et al. [11] proposed a hole bypassing technique in a WSN, deployed in forest or mountain, for avoiding hole(s) arising due to fire. Yu et al. [12] proposed a contour model-based coverage hole detection technique and thereafter, they presented a swarm optimization based hole(s) recovery algorithm to eliminate effects of the hole(s) on WSN operations. Zhuang et al. [13] proposed an algorithm based on a cuckoo search to repair coverage hole in WSN and ensure coverage performance of the network. Soundarya et al. [14] and Lao et al. [15] also proposed hole detection and healing mechanism to improve network performance in the presence of hole(s).

All the above discussion considered hole(s) as static and permanent, which is created at the time of deployment, later due to some node failure, damage or power depletion. But, in this paper, the hole is created, increased or decreased dynamically due to the arise of the dumb node(s). Here dumb nature of a node is varying i.e., dumb nature and well behave nature of a node may interchange during its operations.

OBJECTIVE

It is important to cover the total area by the sensors for the proper functioning of a WSN. If there is any hole in the network then it degrades the performance and the accuracy of the WSN. In the presence of a dumb node, the coverage problem gets extra attention due to dynamically increasing or decreasing the hole region in the covered area. Here, the hole region may generate by the dumb node because a dumb node is considered to sense objects in a area but cannot communicate the sensed information to other nodes. This dumb behavior of a node may vary from time to time. Hole region may create, increase or decrease dynamically. In this paper, we have tried to cover the dynamically created hole area by activating a minimum number of sensor node(s) from sleep mode. If there is only one dumb sensor node then it creates the hole of circular shape. If a group of neighboring sensor nodes exhibits dumb nature, then the created hole region (s) is of irregular shape. The dumb nature of a sensor node may vary from time to time. i.e., a sensor node may exhibit dumb nature, but, in the future, it may not and vice versa. The size of the hole depends upon the number of neighboring dumb nodes present in the group. As the dumb nature varies from time to time, the hole area may increase or decrease. The same hole decreases in size or diminishes if some or all the affected nodes start to behave normally. There may be some nodes on the proximity of the hole area, which is in sleep mode. As well as, there may be few nodes on proximity of a hole area whose communication areas are partially or fully overlapping it. These nodes may be in sleep mode. If all these sleep nodes are activated, then there is a chance of redundant coverage (overlapping communication area) of the hole area. Here, our intension is to activate a minimum number of nodes to cover the total hole area, optimally. In this chapter, the optimization technique is used to identify and activate a minimum number of sleep nodes on the proximity of a hole for covering the total hole area. As the size of the hole is dynamic, the selection of sleep node (s) for activation also needs to be dynamic. If all nodes on the proximity of the hole are activated to cover it then the overall power consumption of the network increases. So, here, optimization is used to reduce the power consumption by activating a minimum number of nodes to cover the whole region.

METHODOLOGY

Hole Area Calculation

We assumed that each sensor nodes are GPS enabled. The Center position of each node is known by GPS. The communication range for a sensor node is circular and radius is also known. If the hole area is for one dumb node then it is very easy to find out the area of the whole region. Hole(s) size arises due to more than one collocating sensor nodes are exhibiting dumb nature. If more than one collocating sensor nodes show dumb nature, then an irregular shaped hole is created. The area of this hole is needed to calculate to cover it.

We have seen that all existing procedures for hole area calculation mentioned in literature have calculated approximate hole area rather finding accurate hole area. Here, we had calculated an accurate hole area by using a simple approach. First, we find the intersecting points of the circular communication range between any two collocating sensor nodes. Thereafter, we check, if anyone of the intersecting points is under the communication area of any other collocated sensor node. If anyone of the intersecting points is under the communication area of another dumb sensor node then it is considered as the inner points of the hole and the remaining intersecting point is the outer intersecting point. In this way, we can calculate the outer points of the hole area by finding intersecting points of two collocating sensor nodes.

Let there be five collocated dumb sensor nodes A, B, C, D, E. P_1, and P_2 are two intersecting points between the communication boundary of two collocated dumb nodes as shown in Figure 1. Point P_2 is under the area of the communication range of dumb node D. Similarly, between communication range of nodes B and C, P_3 and P_4 are two intersecting points and P_4 under the area of E. Here P_2 and P_4 are considered as inner intersecting points. Now the question is how we know the intersecting point is under the area of any other circle?

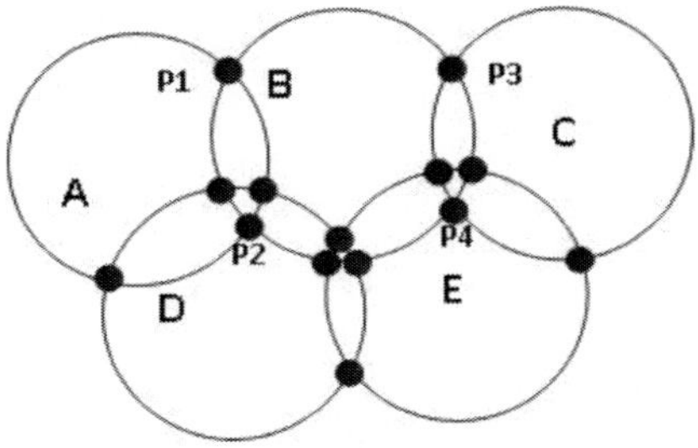

Figure 1. Collocated sensor nodes and their communication boundary.

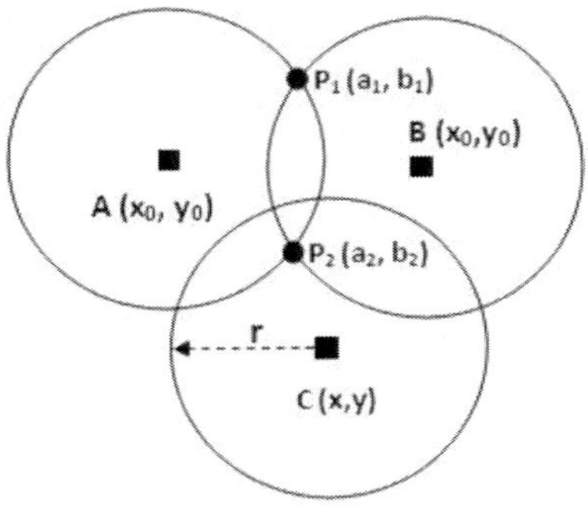

Figure 2. Interior or exterior points.

Let, there be three dumb nodes A, B and C. Their communication range is *r*. Three circles are drawn with the radius of this communication range as shown in Figure 2. The coordinate of the center of circles A, B and C are (x_0, y_0), (x_1, y_1) and (x, y), respectively. These coordinates are measured by the GPS inside the nodes. The total area covered by the dumb nodes is the hole area. For this, we need to find out intersecting points (co-ordinates) between two circles. Thereafter, we can differentiate between inner and outer intersecting points. Here $P_1(a_1, b_1)$ and $P_2\ (a_2, b_2)$ are two intersecting points between two sensor node's communication range. We are calculating intersecting points as follows.

$$a_1 = \frac{(x_1 + x_0)}{2} + \frac{(y_0 - y_1)\sqrt{4r^2 - (x_1 - x_0)^2 - (y_1 - y_0)^2}}{2\sqrt{(x_1 - x_0)^2 + (y_1 - y_0)}} \tag{1}$$

$$a_2 = \frac{(x_1 + x_0)}{2} - \frac{(y_0 - y_1)\sqrt{4r^2 - (x_1 - x_0)^2 - (y_1 - y_0)^2}}{2\sqrt{(x_1 - x_0)^2 + (y_1 - y_0)}} \tag{2}$$

$$b_1 = \frac{(y_1 + y_0)}{2} + \frac{(x_1 - x_0)\sqrt{4r^2 - (x_1 - x_0)^2 - (y_1 - y_0)^2}}{2\sqrt{(x_1 - x_0)^2 + (y_1 - y_0)}} \tag{3}$$

$$b_2 = \frac{(y_1 + y_0)}{2} - \frac{(x_1 - x_0)\sqrt{4r^2 - (x_1 - x_0)^2 - (y_1 - y_0)^2}}{2\sqrt{(x_1 - x_0)^2 + (y_1 - y_0)}} \tag{4}$$

Here, we are getting two coordinates P1 (a1, b1) and P2 (a2, b2). The coordinate P1 or P2 is considered as the inner intersecting point if it belongs to the communication area of node C. If the Euclidian distance between the intersecting point and the center of the third circle C is less than the radius of the circle, then the intersecting point is considered as the inner intersecting point.

Mathematically,

$$\sqrt{(a_1 - x)^2 + (b_1 - y)^2} \leq r \tag{5}$$

or

$$\sqrt{(a_2 - x)^2 + (b_2 - y)^2} \leq r \tag{6}$$

If P1 (a1, b1) satisfy Eqn (v), is in the circle C or if P2 (a2, b2) satisfy Eqn (vi) is in the circle C. After finding all outer intersection points (co-ordinates), we need to arrange these points in an anti-clockwise direction.

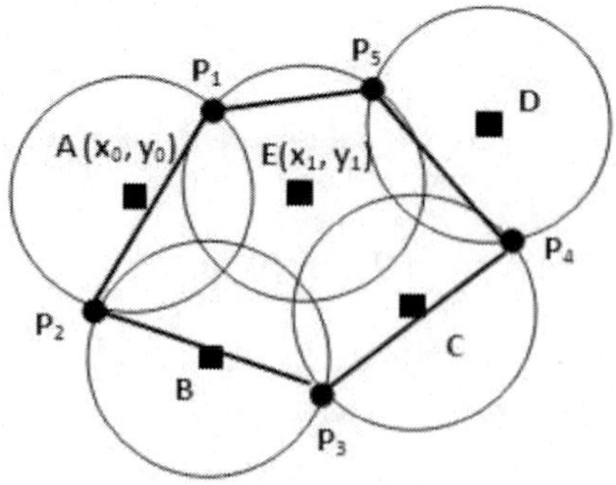

Figure 3. Area of a polygon.

Let, A, B, C, D and E are five dumb nodes as shown in Figure 3. Circular shape indicates the communication range of each node. Using the above Eqns (1-4) we find the intersecting points among all circles. Outer intersecting points P_1, P_2, P_3, P_4 and P_5 are found by the Eqns (5) and (6). After getting all outer intersecting points, we arrange these points in a clockwise or anti-clockwise fashion. Each outer intersecting point is joined with the next coordinate according to the arrangement. After joining them in a polygon shape, we need to calculate the area of the polygon. We calculate area ($\boldsymbol{a_p}$) of the polygon as follows:

$$a_p = \sum_{i=0}^{k-1} \sum_{j=k-1}^{0} (x_j + x_i) \times (y_j - y_i). \tag{7}$$

where k is the number of intersecting points on the contour of the polygon

Thus, we calculate the area of the polygon. Now, we calculate remaining areas i.e., area of sectors of the circles outside the polygon. For this, we determine the sectors are inside or outside of the polygon.

First, we check, if the center of the circle is inside or outside the polygon by using the following condition.

Center of the circle is outside the polygon if,

$$\boldsymbol{If}\ \boldsymbol{x} < MIN(\boldsymbol{X})\ \boldsymbol{OR}\ \boldsymbol{y} < MIN(\boldsymbol{Y})\ \boldsymbol{OR}\ \boldsymbol{x} > MAX(\boldsymbol{X})\ \boldsymbol{OR}\ \boldsymbol{y} > MAX(\boldsymbol{Y})$$

Otherwise, the center is inside the polygon. Here, MIN(X) and MAX(X) are the minimum and maximum x coordinates and similarly, MIN(Y) and MAX(Y) are for minimum and maximum y coordinates. (x,y) is the position of a co-located dumb sensor nodes.

Finding Sector of a Circle

A circle is shown in Figure 4 and O is its center. OA = OB = **r** is the radius, and AB is the chord having length d such that AX = XB = **d/2**. Therefore, $\angle$AOB = **θ**, $\angle$AOX = $\angle$XOB = **θ/2**

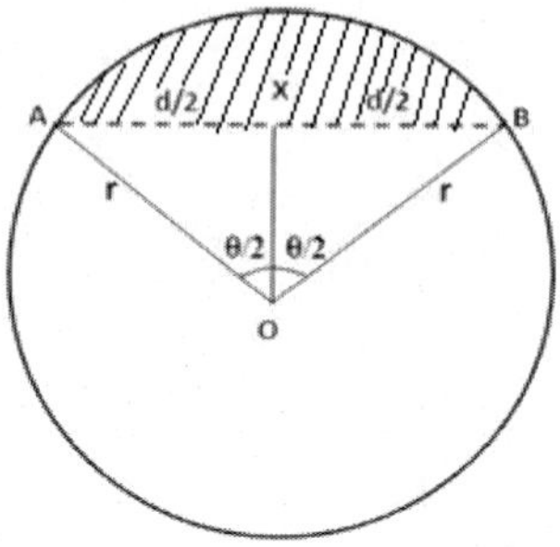

Figure 4. Sector area of a circle.

Now, we calculate the area of shaded region $\boldsymbol{a_i}$
Area of a sector of a circle is:

$$a_i = \frac{r^2}{2}(\theta - \sin\theta)\,. \tag{8}$$

Again,

$$Sin\theta = \frac{BX}{OB}$$

In Figure 4,

$$Sin\theta/2 = \frac{d/2}{r}$$

$$\frac{\theta}{2} = \sin^{-1}\frac{d}{2r}$$

$$\theta = 2\sin^{-1}\frac{d}{2r} \tag{9}$$

Putting the value of $\boldsymbol{\theta}$ in Eqn (viii), we find:

$$a_i = \frac{r^2}{2}(2\sin^{-1}\frac{d}{2r} - \sin(2\sin^{-1}\frac{d}{2r})$$
$$= \frac{r^2}{2}(2\sin^{-1}\frac{d}{2r} - 2\sin(\sin^{-1}\frac{d}{2r})\cos(\sin^{-1}\frac{d}{2r})$$

$$= \frac{r^2}{2}(2 \sin^{-1}\frac{d}{2r} - 2\frac{d}{2r}(\frac{\pi}{2} - \sin\sin^{-1}\frac{d}{2r}))$$

$$\frac{r^2}{2}\left(2 \sin^{-1}\frac{d}{2r} - \frac{d}{r}(\frac{\pi}{2} - \frac{d}{2r})\right)$$

$$a_i = \frac{r^2}{2}\left(2\sin^{-1}\frac{d}{2r} - \frac{d}{r}(\frac{\pi}{2} - \frac{d}{2r})\right) \quad (10)$$

If the center of the communication range of a sensor node is in the outside area of the polygon then we will use

$$a_i = \frac{r2}{2}\left(2\sin^{-1}\frac{d}{2r} - \frac{d}{r}(\frac{\pi}{2} - \frac{d}{2r})\right) \quad (11)$$

and, if the center of the communication range of a sensor node is in the inside area of the polygon then we need to subtract the sector area from the total area of the circle. We can calculate by using the following formula:

$$a_i = \Pi r^2 - \frac{r2}{2}\left(2\sin^{-1}\frac{d}{2r} - \frac{d}{r}(\frac{\pi}{2} - \frac{d}{2r})\right) \quad (12)$$

The total area of all sectors outside the polygon is:

$$a_s = \sum_{i=0}^{n} a_i. \quad (13)$$

Here, we are getting area in two parts, one is for polygon a_p and another is for the sum of the area of all sector a_s.

So, the total hole area is $\mathbf{A} = a_p + a_s$ (14)

Hole Coverage

After estimating the created hole area, we need to cover the hole area by activating the sleep sensor nodes on the proximity of the created hole. Activating all the sleeping sensor nodes on the proximity of the created hole may be a solution for covering the hole area. But this is not a good solution because of redundant coverage due to overlapping communication area of sensor nodes. Here, sensor nodes on the proximity of the hole area are the nodes whose communication range is covering partially or fully the hole area. Mathematically, we can define the sensor node on the proximity of the hole as follows.

$$x > (MIN(X) - r) OR\ y > (MIN(X) - r)$$

$$OR\ x < (MAX(X) + r) OR\ y < (MAX(Y) + r)$$

and

$$A \cap A_s \neq \varphi$$

Here **r** is the communication radius of a sensor node, **A** is the hole area to be covered and $\mathbf{A_s}$ is the communication range of the sleeping sensor node.

Unnecessary activation of sensor nodes reduces the overall lifetime of the sensor network. So, we need to activate some of the sleeping sensor nodes judiciously such that hole area is covered and overlapping between the communication areas of the sensor nodes should be minimum. Hence, the overall power consumption of the network reduces and network lifetime increases.

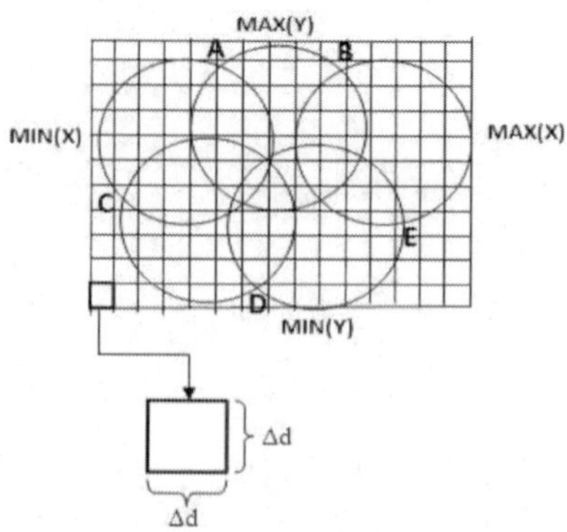

Figure 5. Hole coverage.

Let, there be a whole region. A, B, C, D and E are the intersecting points as well as points on the contour of the region as shown in Figure 5. The hole area is divided into a grid. Each unit square block having sides of Δd, where Δd is a very small constant. The hole coverage scheme scans from *(MIN(X), MIN(Y))* co-ordinate to *(MAX(X), MAX(Y))* co-ordinate row-wise and find out unit square blocks, which are inside the hole. If the hole boundary intersects any block on a row, the scheme starts counting the blocks up to the block in the same row, which intersects the hole boundary. This scanning process is performed by the scheme for each row within *(MIN(X), MIN(Y))* and *(MAX(X), MAX(Y))* to find out the blocks inside the hole. The target of the scheme is to cover these blocks by activating an optimal number of sleeting sensor nodes. Now this coverage problem can easily be mapped into a target coverage problem.

In existing literature [4], target coverage problems considered as an NP-complete problem, so it has no direct solution in polynomial time. We have proposed a greedy heuristic algorithm named *Dynamic Hole Coverage for Dumb Node (DHCDN)* for optimally covering the hole area, which creates, increases or decreases dynamically due to dumb behavior of sensor nodes.

The centroid of blocks within the holes is kept into a list named BLOCK. The scheme initially activates randomly a node from the set of nodes which are in the r distance on the y-axis (MIN(X), MIN(Y)), which covers the hole area partially by its communication range. Then it starts finding another node on the same row within (MAX(X)-r) distance. In case, there is no such node that is in r distance from (MIN(X), MIN(Y)) then it searches for nodes, which are within r distance from the scanned row on the y-axis. Among all the nodes on a row, the scheme activates only the node, which covers the maximum number of uncovered blocks. The entire process continues till it reaches ((MAX(X)-r), MAX(Y)-r)).

As the irregular shape of the hole is created dynamically due to the co-located dumb node, so the scheme needs to run periodically after a certain time interval. This time interval depends upon the available resource in the network.

State Transition Diagram

Generally, a node has active and sleep states. The node switch from the active state to the dumb state when it exhibits dumb behavior as shown in Figure 6. A sleep node switches to an active state when it covers the maximum number of uncovered blocks within a hole area. A node changes its state from active to sleep if it is not selected after a time interval **Δt.** The scheme repeats this state transition after every Δt time.

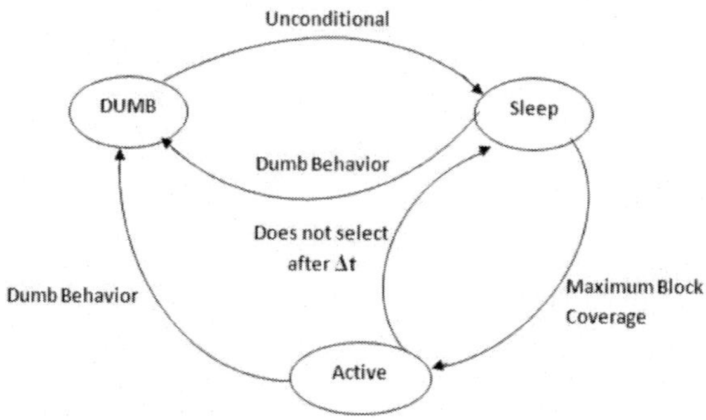

Figure 6. State transition diagram.

Algorithm

Input:

- BLOCK: Centroid of unit blocks
- (MIN(X), MIN(Y)): Coordinate of starting point of hole area
- (MAX(X), MAX(Y)): Coordinate of ending point of hole area

Output:

- block (t_i): List of nodes optimally covering the hole area

1. Start
2. Repeat 2-19 after every Δt time // For Dynamic Hole Occurrence

/* Scan and add blocks within hole area */

3. Divide (MIN(X), MIN(Y), MAX(X), MIN(Y)) into grid of Δd

4. i ← MIN(X)
5. j ← MIN(Y)
6. while (j≤MAX(Y))
7. while (i≤MAX(X))
8. Add centroid of blocks within two consecutive cut into-array BLOCK
9. i++
10. end while
11. j++
12. end while

/*Finding nodes for optimal coverage of hole area*/

13. k ← MIN(Y) + r
14. l ← MIN(X)
15. while(k ≤ MAX(Y))
16. while(l≤MAX(X))
17. Choose a node covering maximum uncovered area-block (t_i) ← BLOCK
18. BLOCK= BLOCK - t_i
19. l = current_position + r
20. end while
21. k ← k+r
22. end while
23. End

CONCLUSION

In this chapter, we have proposed the *DHCDN* scheme to cover dynamically created communication hole(s) area arise due to the dumb behavior of sensor nodes in a WSN by activating the optimal number of sensor nodes. The proposed scheme eliminates dynamic hole(s) from the WSN in an energy-efficient way. The scheme reduces data loss in the network and improves overall network performance. It works in a distributed manner, which in turn increases the reliability of the scheme as well as the overall network. However, the scheme may incorporate some computational overhead on the sensor nodes, which may result in degraded performance and lifetime of individual nodes.

REFERENCES

[1] Akyildiz, I. F.; Weilian Su; Sankarasubramaniam, Y.; Cayirci, E.;, "A survey on sensor networks," *Communications Magazine, IEEE*, vol. 40, no. 8, pp. 102- 114, Aug 2002.

[2] F. Akyildiz, W. Su, Y. Sankarasubramaniam, and E. Cayirci. 2002. "Wireless sensor networks: a survey," *Comput. Netw.* vol. 38, no. 4, pp. 393-422, March 2002.

[3] Wint Yi Poe and Jens B. Schmitt. 2009. "Node deployment in large wireless sensor networks: coverage, energy consumption, and worst-case delay," In *Asian Internet Engineering Conference* (AINTEC '09). ACM, New York, NY, USA, 77-84.

[4] Cardei, M.; Thai, M. T.; Yingshu Li; Weili Wu; "Energy-efficient target coverage in wireless sensor networks," INFOCOM 2005. 24th Annual Joint Conference of the IEEE Computer and Communications Societies. *Proceedings IEEE*, vol. 3, no., pp. 1976-1984 vol. 3, 13-17 March 2005 doi: 10.1109/INFCOM.2005. 1498475.

[5] Xiangai Liu; Yanli Feng; Qingcong Lv; Tongtong Zhao; "Cascaded Movement Strategy for Repairing Coverage Holes in Wireless Sensor Networks," *Information Technology, Computer Engineering and Management Sciences (ICM), 2011 International Conference on*, vol. 2, no., pp. 108-111, Sept. 2011.

[6] Xiao-Heng Deng; Chu-Gui Xu; Fu-Yao Zhao; Yi Liu; "Repair Policies of Coverage Holes Based Dynamic Node Activation in Wireless Sensor Networks," Embedded and Ubiquitous Computing (EUC), *2010 IEEE/IFIP 8th International Conference on*, vol., no., pp. 368-371, 11-13 Dec. 2010.

[7] Weijia Jia; Tian Wang; Guojun Wang; Minyi Guo, "Hole Avoiding in Advance Routing in Wireless Sensor Networks," *Wireless Communications and Networking Conference, 2007. WCNC 2007. IEEE*, vol., no., pp. 3519-3523, 11-15 March 2007.

[8] Qing Fang, Jie Gao, and Leonidas J. Guibas. "Locating and bypassing holes in sensor networks," *Mob. Netw. Appl.* vol. 11, no. 2, pp. 187-200, April 2006.

[9] Honghai Zhang and Jennifer C. Hou., *"Maintaining Sensing Coverage and Connectivity in Large Sensor Networks, Ad Hoc & Sensor Wireless Networks,"* Vol. 1, pp. 89–124.

[10] Kun-Ying Hsieh; Jang-Ping Sheu;, "Hole detection and boundary recognition in wireless sensor networks," *Personal, Indoor and Mobile Radio Communications, 2009 IEEE 20th International Symposium on*, vol., no., pp. 72-76, 13-16 Sept. 2009.

[11] S. Wan and Y. Zhang, *"Coverage Hole Bypassing in Wireless Sensor Networks,"* 2016 *12th International Conference on Mobile Ad-Hoc and Sensor Networks (MSN)*, Hefei, China, 2016, pp. 409-411. doi: 10.1109/MSN.2016.075.

[12] X. Yu, M. Xu, L. Cheng and N. Hu, "A novel coverage holes detection and holes recovery algorithm in wireless sensor networks," *The 27th Chinese Control and Decision Conference (2015 CCDC)*, Qingdao, China, 2015, pp. 3640-3644. doi: 10.1109/CCDC. 2015.7162556.

[13] Y. Zhuang, C. Wu and Y. Zhang, "A Coverage Hole Recovery Algorithm for Wireless Sensor Networks Based on Cuckoo Search," 2017 *IEEE 7th Annual International Conference on CYBER Technology in Automation, Control, and Intelligent Systems* (CYBER), Honolulu, HI, USA, 2017, pp. 1552-1557. doi: 10.1109/ CYBER.2017.8446376.

[14] Soundarya and V. Santhi, "An efficient algorithm for coverage hole detection and healing in wireless sensor networks," *2017 1st International Conference on Electronics, Materials Engineering and Nano-Technology (IEMENTech)*, Kolkata, India, 2017, pp. 1-5. doi: 10.1109/IEMENTECH.2017.8076995.

[15] H. Y. Lao and F. Ding Yi, "Coverage Hole Detection Algorithm Based on HPNs in WSN," *2018 IEEE 18th International Conference on Communication Technology (ICCT),* Chongqing, China, 2018, pp. 896-900. doi: 10.1109/ICCT.2018.8600223.

About the Editors

Dr. G. G. Md. Nawaz Ali

Assistant Professor
Department of Applied Computer Science
University of Charleston, WV 25304, US
Email: ggmdnawazali@ucwv.edu

Dr. Md. Noor-A-Rahim

Senior Post-Doctoral Researcher (MSCA Fellow)
School of Computer Science and IT
University College Cork, Ireland
Email: m.rahim@cs.ucc.ie.

Dr. Pushpendu Kar

Assistant Professor, ERCIM Fellow
School of Computer Science
University of Nottingham Ningbo China,
Ningbo, China
Email: pushpendu.kar@nottingham.edu.cn.

INDEX

A

accelerometers, 133
access, 4, 7, 8, 28, 64, 67, 70, 83, 85, 89, 93, 95, 100, 101, 162, 177, 189, 197, 209, 230
ad hoc network, vii, 25, 26, 28, 29, 30, 31, 32, 107, 109, 112, 114, 136, 173, 174, 177, 178, 179, 180, 181, 183, 184, 185
ad hoc routing, 30, 31
algorithm, ix, x, 16, 30, 33, 35, 36, 37, 38, 39, 60, 61, 64, 66, 67, 68, 69, 70, 71, 74, 75, 76, 77, 78, 82, 85, 86, 89, 93, 95, 102, 105, 106, 110, 122, 125, 126, 132, 133, 137, 138, 139, 140, 141, 158, 159, 160, 161, 162, 163, 175, 176, 179, 181, 184, 193, 197, 198, 203, 204, 212, 215, 216, 231, 242, 251, 255
applications, v, vi, vii, viii, ix, xi, 1, 2, 3, 4, 5, 8, 25, 26, 28, 29, 33, 34, 35, 36, 42, 45, 61, 64, 67, 68, 109, 110, 112, 118, 173, 174, 177, 180, 181, 183, 187, 188, 189, 192, 197, 208, 210, 216, 220, 225, 226, 227, 229, 230, 232, 234, 235
architecture, x, 7, 8, 28, 34, 36, 41, 42, 43, 47, 52, 53, 54, 56, 57, 58, 59, 60, 62, 70, 72, 84, 133, 134, 135, 136, 140, 175, 176, 179, 188, 191, 192, 222, 227, 231, 236
arithmetic, 40, 42, 144
asynchronous communication, 136, 169, 171
attacker, 195, 200, 203, 225
authentication, 28, 37, 62, 132, 197
automate, 117, 124, 143, 223
automation, 217

B

bandwidth, x, 13, 17, 18, 19, 64, 65, 71, 73, 74, 82, 85, 86, 97, 106, 116, 129, 130, 146, 163, 222, 223
bitcoin, 188, 194, 204, 208, 211, 212, 213, 214, 215, 216, 218, 225, 228, 232, 233, 234, 237
black hole, 131, 178
blockchain, vi, viii, xi, 187, 188, 189, 190, 191, 192, 193, 194, 195, 196, 197, 198, 199, 200, 201, 202, 203, 204, 205, 206, 207, 208, 209, 210, 211, 213, 214, 216, 217, 220, 221, 222, 223, 224, 225, 226,

227, 228, 229, 230, 231, 232, 234, 235, 236, 237, 238

C

challenges, vii, viii, xi, 25, 26, 28, 67, 115, 116, 140, 173, 188, 192, 197, 222, 225, 227, 228, 234, 235, 236
classification, 3, 106, 144
coding, viii, x, 63, 64, 65, 66, 67, 69, 70, 71, 73, 74, 75, 76, 77, 78, 79, 80, 81, 82, 83, 84, 86, 87, 88, 89, 93, 94, 95, 96, 97, 99, 100, 102, 103, 105, 106, 107, 108, 109, 111, 112, 171, 185
communication, vii, viii, ix, xii, 1, 2, 4, 5, 6, 7, 8, 13, 15, 19, 20, 24, 28, 29, 31, 33, 34, 37, 61, 64, 70, 71, 106, 114, 129, 131, 133, 136, 142, 146, 148, 150, 152, 153, 154, 155, 156, 159, 161, 162, 163, 165, 166, 167, 168, 170, 171, 179, 185, 197, 225, 239, 240, 241, 242, 243, 244, 245, 246, 247, 249, 250, 251, 253
communication hole, xii, 239, 240, 241, 253
communication overhead, 19, 131, 133, 142, 155, 163, 171, 242
communication systems, 28, 31
communication technologies, 64
complexity, x, 41, 42, 56, 64, 68, 75, 78, 111, 135, 141, 142, 143, 146, 150, 152, 153, 156, 161, 162, 163, 165, 166, 170, 204
computation, 2, 7, 35, 36, 40, 41, 42, 45, 46, 59, 60, 69, 71, 131, 222
computer, 14, 26, 29, 174, 207, 211, 213
computing, 30, 31, 44, 64, 110, 174, 175, 176, 177, 178, 184, 216, 217, 221, 222, 236
connectivity, 5, 9, 17, 19, 22, 24, 31, 115, 122, 152, 162, 198, 242
consensus, xi, 188, 189, 190, 191, 192, 196, 198, 199, 200, 201, 202, 203, 204, 207, 209, 211, 212, 215, 216, 221, 222, 223, 225, 226, 228, 231, 234
consensus algorithms, 188, 192, 196, 198, 203, 204, 212, 226, 228
construction, 71, 77, 157, 213
correlation, 18, 140, 182
coverage, vi, xii, 131, 239, 240, 241, 243, 249, 250, 251, 253, 254, 255, 256
CPU, 35, 115, 134, 200, 212, 216
cryptocurrencies, 188, 192, 209, 211, 214, 216, 218, 228, 231, 232, 233
cryptography, 34, 37, 38, 62, 189, 213, 228

D

data broadcasting, 64, 107, 111
data center, 34, 220
data distribution, 221
data processing, 221, 228
data structure, 132, 192, 194
data transfer, 121, 216, 221
database, 64, 71, 72, 84, 85, 86, 89, 190, 194, 198, 209, 221
decoding, 59, 65, 69, 71, 108, 171
dependability, x, 113, 114, 115, 118, 119, 120, 121, 123, 143, 172, 174, 175, 181
detection, 4, 26, 27, 123, 131, 132, 133, 134, 135, 136, 137, 138, 139, 142, 158, 174, 178, 179, 180, 181, 222, 242, 255
detection techniques, 123, 135
distributed computing, 176, 183
dumb node, xii, 240, 241, 242, 243, 244, 245, 247, 251

E

elliptic curve cryptography, v, viii, ix, 33, 34, 35, 37, 38, 60, 61, 62
emergency, 4, 27, 36, 64
encoding, 65, 70, 71, 73, 75, 88, 106, 112
encryption, 37, 38, 62, 190, 196, 197

end-users, viii, 1
energy, xi, 5, 126, 131, 134, 135, 142, 152, 162, 163, 168, 169, 179, 194, 201, 202, 211, 212, 213, 220, 231, 234, 239, 240, 241, 253, 254

F

fault detection, 123, 130, 132, 134, 137, 139, 140, 164, 173, 179, 180, 181, 182, 183
fault diagnosis, viii, xi, 26, 113, 114, 116, 117, 118, 122, 123, 139, 140, 141, 143, 146, 147, 149, 150, 154, 155, 156, 158, 160, 161, 169, 170, 172, 174, 175, 176, 177, 179, 182, 183, 184
fault tolerance, 121, 122, 137, 143, 172, 201, 234
fault tolerant, 114, 175
field programmable gate arrays (FPGA), 34, 35, 36, 41, 61, 62

G

gateway node, 158
governance, 227, 228
graph, x, 63, 66, 70, 71, 72, 73, 74, 75, 76, 77, 78, 82, 106, 124, 125, 126, 127, 141, 146, 152, 159, 169, 177
grids, 32, 231

H

hacking, 227
healing, 133, 178, 242, 255
hybrid, 10, 11, 14, 15, 19, 20, 23, 25, 30, 109, 174, 178, 200
hybrid routing, 10, 11, 19, 23

I

infrastructure, vii, 4, 14, 37, 107, 115
integration, 70, 191, 222, 227, 230, 231
intelligent transport systems (ITS), v, vii, viii, ix, 1, 2, 4, 26, 33, 34, 35, 36, 37, 40, 42, 45, 61
intermediaries, xi, 188, 190, 224
interoperability, 189, 227
IP address, 221
issues, viii, 14, 21, 118, 151, 174, 181, 191, 220, 221, 222, 225, 226, 228

L

languages, 227
latency, ix, 10, 33, 35, 42, 44, 52, 54, 56, 57, 59, 114, 126, 136, 137, 145, 149, 150, 152, 154, 155, 157, 160, 161, 167, 168, 169, 208, 222, 223
layered architecture, 7, 8

M

MANET(s), viii, x, 1, 2, 4, 5, 6, 9, 25, 30, 113, 114, 130, 131, 133, 136, 138, 140, 149, 150, 151, 152, 153, 155, 156, 157, 158, 159, 161, 164, 166, 178, 181, 182, 183, 184
memory, x, 6, 10, 18, 34, 41, 42, 56, 58, 59, 115, 121, 134, 144, 199, 204, 211, 216
mesh networks, 69, 111, 114, 136, 173, 180, 185
messages, ix, 18, 34, 35, 43, 60, 116, 132, 133, 135, 136, 137, 142, 146, 148, 149, 151, 152, 153, 155, 156, 158, 160, 162, 163, 166, 168, 169
mobile communication, 134

mobile wireless networks, viii, x, 113, 114, 115, 116, 118, 141, 143, 146, 167, 171, 172, 173, 174, 184
models, 5, 67, 71, 112, 118, 124, 125, 126, 127, 128, 129, 130, 133, 140, 141, 143, 146, 172, 177
modules, 34, 58, 59
multiple nodes, 190, 207
multiplication, ix, 34, 35, 39, 40, 41, 42, 43, 44, 45, 46, 49, 56, 57, 58, 59, 60, 61
multiplier, 40, 41, 42, 45, 49, 50, 52, 53, 57, 62

N

navigation system, 64
network coding, x, 64, 65, 66, 67, 69, 70, 71, 74, 75, 78, 79, 80, 81, 87, 88, 95, 97, 102, 103, 105, 106, 108, 109, 111, 112, 171
network congestion, 18, 116
networking, 31, 110, 134, 177, 179, 217, 222
neural network, 26
nodes, viii, xi, 1, 2, 5, 6, 7, 8, 9, 10, 11, 17, 18, 19, 24, 26, 61, 107, 112, 114, 115, 116, 117, 122, 124, 125, 126, 128, 129, 130, 131, 132, 133, 134, 135, 136, 137, 138, 139, 142, 143, 144, 145, 146, 147, 149, 150, 151, 152, 153, 154, 155, 156, 157, 158, 159, 160, 161, 163, 164, 166, 167, 168, 169, 170, 171, 172, 174, 178, 179, 190, 196, 197, 198, 199, 201, 202, 203, 206, 207, 212, 213, 215, 221, 222, 226, 239, 240, 241, 243, 244, 245, 247, 249, 250, 251, 252, 253

O

on board unit (OBU), viii, 1, 2, 6, 7
operating system, 115, 121
operations, 38, 70, 114, 116, 126, 144, 172, 224, 242
outsourcing, 223

P

packet forwarding, 19, 20, 22
parallel processing, ix, 34, 40, 41, 42, 60
platform, 35, 208, 221, 223, 224, 227, 235
policy, 71, 85, 112
probability, 44, 89, 116, 119, 120, 127, 129, 145, 201, 202, 213, 215
programmability, 211, 222
programming, 174, 227
propagation, 9, 140, 141, 143, 234
public sector, 209, 210

Q

quality of service, 136, 180
query, 64, 65, 109

R

radio, 3, 6, 15, 18, 22, 223
radius, 244, 245, 246, 247, 250
recovery, 17, 19, 109, 142, 143, 234, 242, 255
recovery process, 143
redundancy, 111, 122, 137, 138, 226
reliability, 19, 35, 109, 119, 121, 225, 253
repair, 119, 176, 242
requirement(s), vii, x, xi, 14, 34, 45, 60, 64, 68, 113, 116, 119, 121, 127, 136, 143, 167, 172, 173, 189, 191, 208, 213, 227
resource management, 220, 222, 228, 235
resources, 35, 115, 117, 151, 163, 166, 199, 215, 220, 222, 223, 225, 226, 234
response, viii, 65, 66, 68, 69, 86, 93, 102, 103, 105, 135, 147, 148, 150, 151, 154,

156, 158, 160, 162, 163, 169, 174, 181, 222
response time, viii, 65, 68, 69, 86, 93, 102, 103, 105, 222
root, 44, 51, 60, 123, 133, 140, 141, 142, 156, 192, 194, 195, 204, 211
rules, 72, 82, 121, 192, 199, 209, 226, 227

S

safety, viii, 2, 4, 8, 13, 25, 26, 35, 36, 65, 69, 108, 109, 119, 222
scale system, 180
scanning process, 251
scarce resources, 116
scheduling, x, 20, 63, 64, 65, 66, 67, 68, 69, 70, 71, 73, 74, 75, 76, 77, 78, 79, 80, 82, 85, 86, 89, 95, 97, 103, 106, 107, 108, 109, 110, 111, 112
security, viii, xi, 7, 26, 30, 34, 35, 36, 37, 38, 45, 59, 60, 177, 188, 189, 190, 191, 198, 209, 211, 213, 215, 217, 220, 221, 222, 225, 226, 228, 230, 234, 235, 236, 238
sensor network, 26, 173, 176, 178, 179, 181, 240, 242, 250, 254, 255
sensor nodes, xi, 139, 174, 178, 239, 240, 241, 243, 244, 245, 247, 249, 250, 251, 253
sensor(s), viii, xi, 1, 2, 3, 4, 26, 34, 126, 134, 137, 138, 139, 173, 174, 176, 178, 179, 181, 217, 239, 240, 241, 242, 243, 244, 245, 247, 249, 250, 251, 253, 254, 255
service provider, 212
services, viii, x, 1, 7, 11, 64, 65, 67, 110, 113, 114, 115, 118, 119, 120, 121, 122, 141, 172, 222
simulation(s), x, 64, 66, 67, 70, 84, 85, 86, 97, 106, 149, 150, 153, 156, 159, 161, 162, 166, 167, 171
smart contract, 188, 191, 196, 208, 209, 210, 211, 217, 223, 224, 227, 228, 231, 232, 238
software, 35, 115, 121, 122, 130, 133, 134, 137, 142, 179, 191, 198, 208, 213, 222, 232, 236, 238
solution, 35, 36, 38, 56, 60, 64, 70, 136, 189, 214, 217, 220, 232, 249, 251
storage, 115, 216, 222, 225, 226, 228, 233, 234, 236
strategy use, 126
structure, 6, 11, 59, 114, 115, 128, 163, 166, 195, 205
subnetworks, 203, 214
supply chain, 191, 208, 217, 223, 224, 232, 237

T

target, 20, 22, 122, 194, 199, 200, 213, 251, 254
taxonomy, 172, 174, 175, 178, 228
techniques, vii, viii, xi, 5, 15, 35, 40, 41, 42, 59, 60, 64, 65, 113, 118, 121, 122, 123, 124, 130, 131, 133, 134, 135, 137, 139, 141, 142, 143, 163, 170, 172, 174, 181, 196, 197, 198, 211, 215, 220, 232
technologies, vii, 2, 190
technology, viii, xi, 32, 34, 179, 188, 189, 190, 192, 202, 208, 209, 210, 211, 216, 217, 220, 221, 222, 223, 224, 225, 227, 230, 237
testing, vii, 124, 125, 126, 127, 134, 147, 149, 150, 151, 153, 156, 160, 162, 164, 167, 170, 171, 179
threats, 118, 121, 172, 175, 222
topology, ix, 2, 5, 11, 17, 22, 25, 26, 115, 116, 118, 122, 129, 131, 136, 146, 149, 150, 151, 153, 155, 157, 158, 159, 161, 162, 163, 164, 166, 168, 169, 170, 171, 173, 183

transactions, xi, 32, 175, 178, 188, 189, 190, 191, 192, 194, 195, 198, 199, 201, 203, 204, 207, 209, 211, 212, 213, 214, 221, 224, 225, 226, 238
transformation, 81, 204
transistor, 134
transport, ix, 33, 34, 36, 61

V

validation, 56, 192, 193, 214, 221
VANET(s), v, viii, 1, 2, 3, 4, 5, 6, 7, 8, 9, 11, 13, 14, 15, 25, 26, 27, 28, 29, 30, 31, 32, 107, 110, 114, 134, 173
variations, 131, 157
vector, 29, 125, 136, 140
vehicles, vii, viii, 1, 2, 3, 4, 6, 7, 9, 12, 13, 15, 17, 24, 25, 34, 35, 37, 65, 134, 157

W

Wi-Fi, 3
wireless devices, 134, 179
wireless LANs, 66
wireless networks, viii, x, 29, 31, 111, 113, 114, 115, 116, 118, 123, 129, 130, 131, 132, 133, 137, 139, 140, 141, 143, 146, 147, 167, 171, 172, 173, 174, 178, 182, 184
wireless sensor network, 26, 61, 114, 174, 176, 179, 180, 181, 182, 240, 254, 255
wireless sensor networks, 61, 114, 174, 176, 179, 180, 181, 182, 254, 255
wireless technology, 3